Brick Layers
Creative Engineering with LEGO® Constructions

Authors

Sheldon Erickson   Tom Seymour
Martin Suey

Editors

Betty Cordel   Sheldon Erickson
Judith Hillen

Illustrator

Reneé Mason

Desktop Publisher

D1210784

Kristy Shuler

This book contains materials developed by the AIMS Education Foundation. **AIMS** (**A**ctivities **I**ntegrating **M**athematics and **S**cience) began in 1981 with a grant from the National Science Foundation. The non-profit AIMS Education Foundation publishes hands-on instructional materials (books and the monthly *AIMS* magazine) that integrate curricular disciplines such as mathematics, science, language arts, and social studies. The Foundation sponsors a national program of professional development through which educators may gain both an understanding of the AIMS philosophy and expertise in teaching by integrated, hands-on methods.

ISBN 1-881431-62-2

Printed in the United States of America

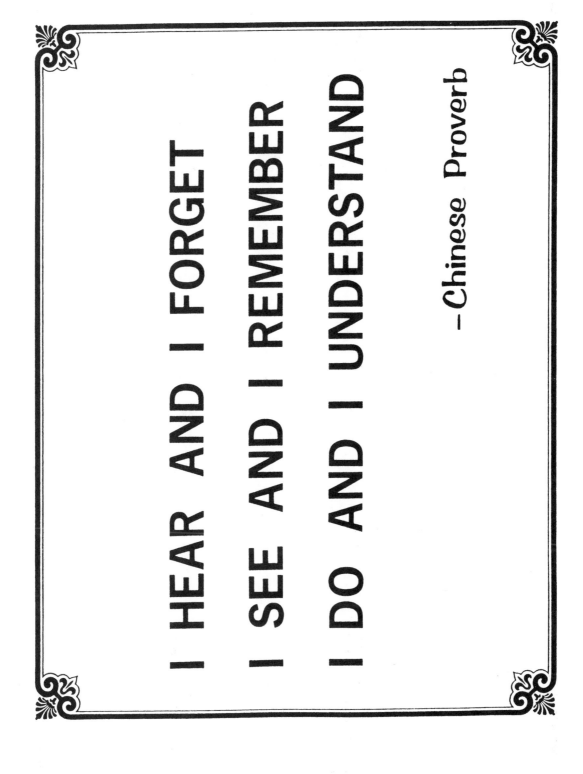

I HEAR AND I FORGET

I SEE AND I REMEMBER

I DO AND I UNDERSTAND

—Chinese Proverb

Brick Layers
Creative Engineering with LEGO® Constructions
Table of Contents

Conceptual Overview

Brick Layers is written to be used with LEGO DACTA® kits. The LEGO® materials provide a natural incentive for student involvement. The relatively easy-to-assemble models provide consistent, reproducible data allowing the methods and processes of science to be practiced. Mathematics is easily integrated into the lessons because the models lend themselves to accurate quantification.

The activities in this book allow students the opportunity to explore the two areas of mechanical and structural engineering. Following is a general overview of the content from each area covered in the book and a visual organization of how they are related.

Mechanical Engineering

The underlying concept of all the activities in this section is the conservation of energy. Machines do not create energy to get work done, they only alter the factors of work: the force used and the distance through which that force moves. This concept is coupled with the understanding that every machine involves a trade-off: If a machine alters a small force to a great force, it requires that small force to go a great distance to make the great force go a small distance.

Following is a list of the supporting content developed in this section.

- Energy cannot be created or destroyed; it can only be changed from one form to another
- Energy is used to accomplish work.
    Work is done when something is moved.
    The amount of work done is determined by how much force was used to move the object and how far the object was moved.
- Friction is a force that resists motion. Friction takes energy away from a machine that could otherwise be used to accomplish useful work.
- Mechanical advantage is a comparison of force produced by a machine to the force applied to the machine.
- A machine is a device that inversely changes the amount of force and distance through which the force is applied.
    Inclined planes are flat, sloping surfaces over which objects can be pushed or pulled to move them to a higher level. Inclined planes can be used to raise objects with less force through a greater distance than a straight lift.
    Levers are bars that pivot on a fulcrum. A lever increases force when the force is placed a greater distance from the fulcrum than the resistance it is moving.
    Wheels and axles are turning levers. The wheel and axle increase force when a small force turns the wheel a long distance resulting in the axle turning a small distance with a great force.
    Gears are specialized wheels and axles with teeth that discourage slippage. The number of teeth on a gear is a measure of the distance around the gear and simplifies the calculation of energy transfer.

v

Structural Engineering

The underlying concept of all the activities in this section is that shapes have special properties that make them useful in construction. The section focuses on the cycloid and the triangle.

Following is a list of the supporting content developed in this section.

- Shapes have special properties that make them useful in specific engineering applications.
- The cycloid is the path traced by a point on the circumference of a circle as it is rolled along a straight line.
- The triangle, with the property of intrinsic stability, is used to develop strength in the construction of structures.
- Tension and compression are forces that are present and need to be considered in building structures.

At the bottom of the pages on the right side of this book, you will find an illustration of a gear on a gear rack. When the pages are thumbed through (flipped), the gear will appear to move along the gear rack. The head lamp of the little figure in the gear will make a cycloidal shape as it appears to move.

After their study of cycloids, students are invited to *Make Your Own Flip Book* using the same illustrations that have appeared throughout this book.

Mechanical Engineering Concepts

Energy cannot be created or destroyed; it can only be changed from one form to another.

Energy is used to accomplish work.

Work is done when something is moved. The amount of work done is determined by how much force was used to move the object and how far the object was moved.

Mechanical advantage is a comparison of force produced by a machine to the force applied to the machine.

Friction is a force that resists motion. It takes energy from a machine that could be used to get useful work accomplished.

A machine is a device that inversely changes the amount of force and the distance through which the force is applied.

Levers are bars that pivot on a fulcrum. A lever increases force when the force is placed a greater distance from the fulcrum than the resistance it is moving.

Inclined planes are flat, sloping surfaces over which objects can be pushed or pulled to move them to a higher level. Inclined planes can be used to raise objects with less force through a greater distance than a straight line.

Wheels and axles are turning levers. The wheel and axle increase force when a small force turns the wheel a long distance resulting in the axle turning a small distance with a great force.

Gears are specialized wheels and axles with teeth that discourage slippage. The number of teeth on a gear is a measure of the distance around the gear and simplifies the calculation of energy transfer.

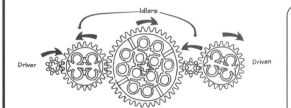

Structural Engineering Concepts

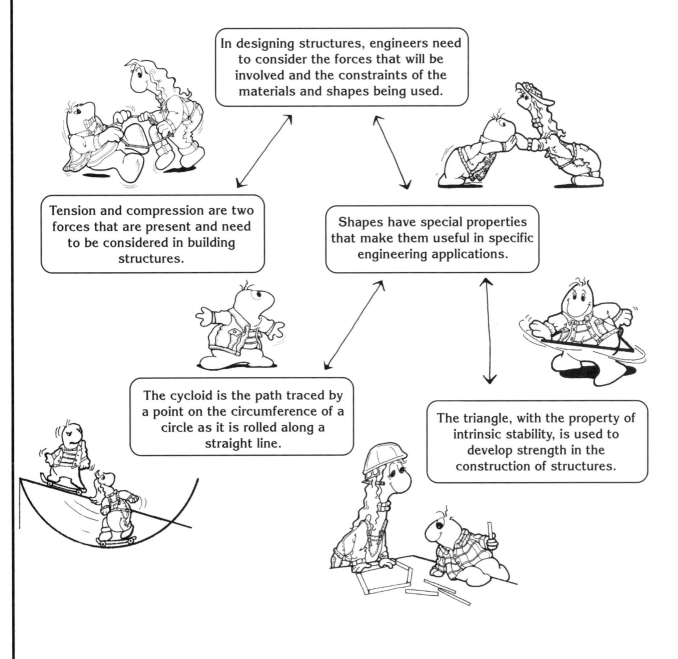

In designing structures, engineers need to consider the forces that will be involved and the constraints of the materials and shapes being used.

Tension and compression are two forces that are present and need to be considered in building structures.

Shapes have special properties that make them useful in specific engineering applications.

The cycloid is the path traced by a point on the circumference of a circle as it is rolled along a straight line.

The triangle, with the property of intrinsic stability, is used to develop strength in the construction of structures.

MATH	Using Computation	Using Rational Numbers	Estimating	Measuring	Graphing	Averaging	Identifying Patterns	Using Geometry & Spacial Sense	Using & Applying Formulae	Using Proportional Reasoning
Force-Ups			x	x	x					
M.V.P. (Most Valuable Place)										
Fiddling with Fulcrums	x		x		x	x			x	
Beams Over Board	x		x						x	
Effort-Less				x	x		x			
Paul Bunyan's Bear-Barrow & Challenge				x			x			
A Balance Beam	x			x			x		x	
Wheeling Your Way to the Top	x			x	x	x	x		x	x
A Shift in Lift	x	x		x	x	x	x			
Slot Cars	x			x	x	x				
LEGO® Launcher	x			x	x	x				
Magic String				x					x	
Meshing Around: Exploring Gear Trains							x			
Reel Changes					x		x			
Turn Around	x						x		x	
Dial - a - Gear		x			x			x		x
The Governor Rules							x			
Gear Guessing				x			x			x
Bug on a Roll								x		
Bug-A-Long				x				x		x
Speed Bugs		x		x	x			x		
A Stable Table								x		
Angle Fixer							x	x	x	
Stress on a String								x		
The Big Boom Construction Project	x			x				x		

INTEGRATED PROCESSES

	Observing	Comparing & Contrasting	Collecting & Recording Data	Interpreting Data	Predicting/ Inferring	Applying/ Generalizing
Force-Ups	X	X	X	X		X
M.V.P. (Most Valuable Place)	X		X	X		X
Fiddling with Fulcrums	X		X	X		X
Beams Over Board	X		X	X		X
Effort-Less	X	X	X	X	X	X
Paul Bunyan's Bear-Barrow & Challenge	X	X	X		X	X
A Balance Beam	X		X	X	X	X
Wheeling Your Way to the Top	X		X	X	X	X
A Shift in Lift	X	X	X	X	X	X
Slot Cars	X	X	X	X		X
LEGO® Launcher	X	X	X	X		X
Magic String	X	X	X	X		X
Meshing Around: Exploring Gear Trains	X	X	X	X		X
Reel Changes	X	X	X	X		X
Turn Around	X	X	X		X	X
Dial - a - Gear	X	X	X	X		X
The Governor Rules	X	X	X	X		X
Gear Guessing	X		X		X	X
Bug on a Roll	X		X			X
Bug-A-Long	X	X	X	X		X
Speed Bugs	X	X	X	X		X
A Stable Table	X	X	X			X
Angle Fixer	X	X	X	X		X
Stress on a String	X	X				X
The Big Boom Construction Project	X	X				X

Classroom Management with LEGO® Materials

Set The Tone

- **This is science equipment**
- **Learn the names of the parts**

Not surprisingly, the natural inclination is to view LEGO® elements as toys. It is important to distinguish between LEGO DACTA® kits used with these activities and the LEGO® toy sets many students may have at home.

LEGO DACTA® kits are developed by LEGO Dacta, the educational division of the LEGO Group, and include features designed specifically for classroom use.
- Building instructions for many different models
- Models selected for curriculum relevance
- A storage tray to assist with inventory and management
- Sufficient quantity for student team use

Treat the LEGO DACTA® kits as science equipment to establish this subtle, yet important distinction for your students.

When students learn about the microscope, they learn the names of the parts and their functions before they use them. So, too, with LEGO® materials. By learning the names of the parts, a mind-set is formed which differentiates these pieces from the toys used at home. Naming the pieces also aids in instruction. When a model calls for a "1 x 8 beam," everyone knows which piece to use. Application: When controlling variables in an experiment, knowing that models are made of identical pieces is very important. A student who describes a model as having "one of those little red things" is not using the precise language of a scientist.

General Terminology for Naming LEGO® Elements

Axles - long black rods used for mounting wheels, forming joints, etc. Axles are named by comparing them to a piece with studs and counting how many studs long they are.

Beams - pieces one stud wide, of varying lengths, with holes in the sides. A beam is named by its dimensions in studs.

Bricks - wider pieces named by the number of studs in the length and width.

Gears - toothed wheels named by the number of teeth around their edges.

Plates - thin, flat pieces named by the number of studs in the length and width.

Studs - the bumps on the top of the LEGO® elements

At home, students are comfortable with dumping out the contents of the box of parts and assembling either from instructions or from their imaginations. That is another difference between these science kits and the home toy kits. These do not get dumped. The list of *Work Rules* reinforces this idea.

Work Rules

1. **No element leaves the kit that is not part of the assigned model.**

2. **Any element out of the kit must be on the model.**

3. **No loose elements are allowed on the desktop!**

These are some common sense rules that emphasize the fact that this is a situation different from playing with LEGO® materials at home. The greatest occasion for lost elements is when they are loose on the desk and roll onto the floor, often without being noticed. If you are using the complete kit, then it is easy to enforce the rules. The idea here is that an element is either **on** the model or **in** the kit.

There are instances, especially with younger students, when it is appropriate to distribute only the parts needed for an investigation, not the entire kit. Using a zipper-type plastic bag with the kit number or other identification on it works very well.

To keep small elements from rolling off the desk, a small piece of terry cloth can be included in the bag. The texture of the cloth stops small pieces, especially bushings and connector pegs, from rolling onto the floor.

Accountability

- **Assign kits to teams**
- **Have folders for accountability**

Kits can be labeled with stickers of different colors or numbers. Each team (teams of two are recommended) is assigned a kit which they will use for the **duration** of the unit. This gives an added incentive to be careful with the many small parts; if a part is lost, it may hamper the ability to complete a future model.

Having a folder with an *Inventory Check List* and an *Inventory List* with the quantity and name of all the parts will help to reinforce that each group is accountable for the contents of the kit. These kits are set up so that it is easy to inventory the parts. A quick glance will let you know if everything is in place. Each group should take an inventory of the kit at the end of each session and sign the *Inventory Check List* and list any parts that are missing, or indicate that no parts are missing. (The space should not be left blank.) This sheet should be initialed by the teacher. If the kits are being used by more than one class, the students will want to verify the last group's inventory by making a quick inventory before beginning. This inventory practice takes a little extra time each period, but if followed faithfully, students quickly become very responsible. The payoff is that very few (if any) parts are lost and the kit will serve in good stead for many years.

Strategies for Use

LEGO® materials provide teachers with the tools to easily follow the recommendations of the *Project 2061 Benchmarks, NCTM Standards,* and other educational research. Indications are that children learn best when they can construct their own knowledge. The teacher's role is to plan an environment for discovery, not to dispense scientific facts for students to verify. The teacher then is a consultant and facilitator. The majority of student learning (and frequently teacher learning) takes place during an activity. The activities in this book are exploratory in nature. It takes careful planning by the teacher to assure quality use of class time. Knowledge of desired outcomes of the activities by the teacher is essential for success.

As a **consultant** you...

- make sure students understand the objective of the activity. That is, *What question are we trying to answer?*

- give only information essential to conducting the investigation, but not too much to spoil the *discovery* in the activity.

- make clear at the beginning any limitations on materials, time, or reporting procedures.

- establish collaborative learning group assignments

As a **facilitator** you...

- supply the necessary LEGO® elements and other materials.

- observe students during the activities, keeping them focused on the *Key Question*.

- assist groups by asking questions as you see necessary.

The process of finding an answer is at least as important, in elementary grades, as the answer itself. Students should not be readily helped out of "predicaments" unless there is no other alternative. The more they can figure out solutions to problems (of process as well as content), the more they will learn.

WORK RULES

1. No element leaves the kit that is not part of the assigned model.

2. Any element out of the kit must be on the model.

3. No loose elements are allowed on the desktop!

LEGO® ELEMENTS KIT

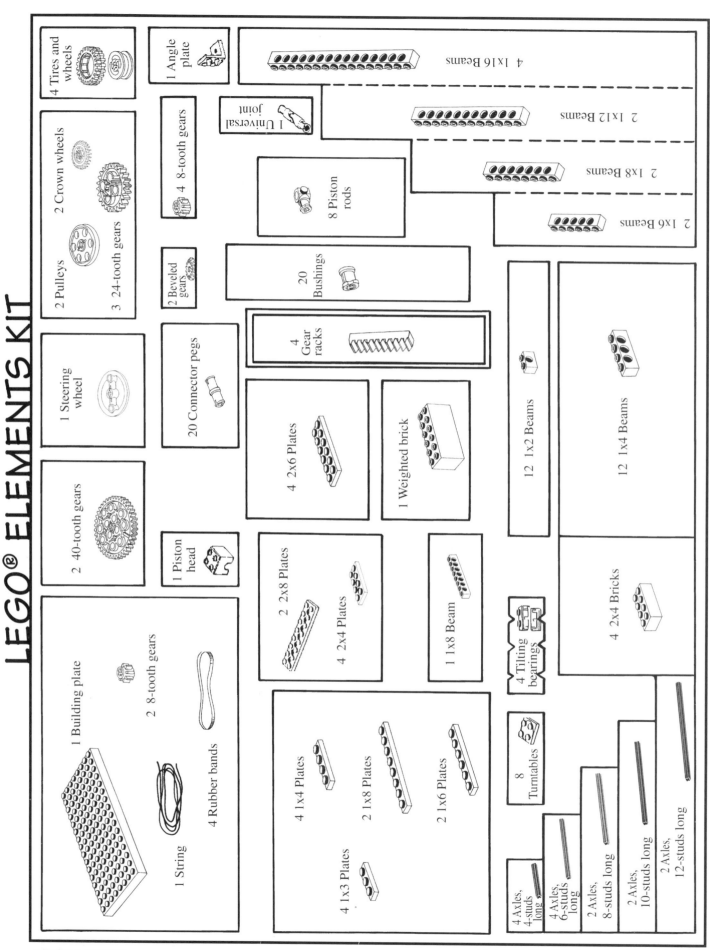

4 Tires and wheels

1 Angle plate

4 1x16 Beams

2 Crown wheels

1 Universal joint

4 8-tooth gears

2 1x12 Beams

2 Pulleys

3 24-tooth gears

8 Piston rods

2 1x8 Beams

2 Beveled gears

20 Bushings

2 1x6 Beams

1 Steering wheel

20 Connector pegs

4 Gear racks

12 1x2 Beams

12 1x4 Beams

2 40-tooth gears

1 Piston head

4 2x6 Plates

1 Weighted brick

2 2x8 Plates

4 2x4 Plates

1 1x8 Beam

4 2x4 Bricks

1 Building plate

2 8-tooth gears

4 Rubber bands

4 1x4 Plates

2 1x8 Plates

2 1x6 Plates

4 Tilting bearings

8 Turntables

1 String

4 1x3 Plates

4 Axles, 4-studs long

4 Axles, 6-studs long

2 Axles, 8-studs long

2 Axles, 10-studs long

2 Axles, 12-studs long

LEGO DACTA® Kit #1030
Inventory List

1	Building plate
4	Rubber bands
1	String
1	Steering wheel
2	Pulleys
2	Crown wheels
4	Tires
4	Wheels
1	Piston head
8	Piston rods
20	Connector pegs
1	Angle plate
4	1 x 3 plates
4	1 x 4 plates
2	1 x 6 plates
2	1 x 8 plates
4	2 x 4 plates
2	Gears, 40-tooth
3	Gears, 24-tooth
6	Gears, 8-tooth
2	Beveled gears
4	Gear racks

2	2 x 8 plates
4	2 x 6 plates
20	Bushings
1	Universal joint
4	Axles, 4-studs long
4	Axles, 6-studs long
2	Axles, 8-studs long
2	Axles, 10-studs long
2	Axles, 12-studs long
8	Turntables
4	Tilting bearings
1	Weighted brick
4	1 x 4 bricks
12	1 x 2 beams
12	1 x 4 beams
2	1 x 6 beams
3	1 x 8 beams
2	1 x 12 beams
4	1 x 16 beams

Inventory Check List

Kit Number_____

Initial Parts Missing_____

_____

Team Members_____

Date	Period	Complete Kit		Notes (missing parts or conditions)
		Student Signature	Teacher Initials	

FORCE-UPS

Topic
Simple Machines: Inclined Planes

Key Question
How does the force needed to pull a weight up a ramp change as the slope of the ramp changes?

Focus
Students will explore the mechanical advantage of a ramp by measuring the change in force required to pull a mass to the top of a ramp. After changing the inclination of the ramp, students will make a generalization that they can apply to determine the mechanical advantage of ramps.

Guiding Documents
NCTM Standards
- *Analyze functional relationships to explain how a change in one quantity results in a change in another*
- *Make inferences and convincing arguments that are based on data*

Project 2061 Benchmarks
- *Energy can change from one form to another, although in the process some energy is always converted to heat. Some systems transform energy with less loss of heat than others.*
- *Graphs can show a variety of possible relationships between two variables. As one variable increases uniformly, the other may do one of the following: always keep the same proportion to the first, increase or decrease steadily, increase or decrease faster and faster, get closer and closer to some limiting value, reach some intermediate maximum or minimum, alternately increase and decrease indefinitely, increase and decrease in steps, or do something different from any of these.*

Math
Estimating
Graphing
Measuring
   angles
   force

Science
Physical science
   simple machines
      inclined planes
      mechanical advantage

Integrated Processes
Observing
Collecting and recording data
Interpreting data
Comparing and contrasting
Generalizing

Materials
LEGO® elements (per group):
1	base plate
2	1 x 16 beams
2	1 x 12 beams
2	1 x 8 beams
2	1 x 6 beams
2	1 x 4 beams
1	2 x 4 brick
1	2 x 4 plate
1	axle, 8-studs long
2	connector pegs
1	bushing
1	weighted brick

Spring scale (250 grams or 2.5 newtons)
String
Ramp (ruler, paint stir stick, etc.)
Tagboard
Tape
35 mm film canister (optional)
Sand (optional)

Background Information
A machine is something that changes the amount or direction of the force used to do work. Some *simple machines* are the inclined plane, lever, wheel and axle, gear, and pulley. Many times these are used in combinations to produce useful *compound machines*.

The answer to "Why do we need to know about simple machines?" is usually found in the mechanical advantage they offer. *Mechanical advantage* is the degree to which a machine changes the effort needed to do a particular task. Mechanical advantage is found by dividing the force of the resistance (the thing being moved) by the force of the effort (the thing doing the moving).

Formulas are used to predict or calculate mechanical advantage. These calculations are for the *ideal mechanical advantage*, what would happen if there were no friction. In reality, however, friction is an element that occurs in every simple machine. If the students' results differ when compared to the ideal formula, friction is usually the culprit.

The inclined plane is the simple machine that is the focus of this activity. An inclined plane may be used to form a ramp, or be put back-to-back as in a wedge, or wrapped around a central point as in a screw. In each case, the amount of effort needed to move the resistance is lessened as the distance through which the effort is exerted is increased.

The most direct way to raise an object four meters is to lift it straight up. This also requires the most effort. A ramp, an inclined plane, allows the object to be pushed up to the four-meter height, but the length of the ramp is greater than four meters. The effort to move the object is less, but it must be exerted over a greater distance.

In the illustration, an object having a mass of six kg needs to be raised to a height of four meters. To lift straight up would require 24 kg•m of work (Work = Force x Distance). Rolling it up the ramp would require rolling twice the distance, but with half the effort. Although the resistance was moved over a greater distance (eight meters), the end result was a four meter increase in elevation from its starting position. Thus the formula would be 24 kg•m of work = 3 kg of effort x 8 m of distance.

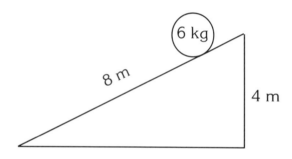

When students calculate mechanical advantage for a ramp, they often find that, in reality, more work is done with the ramp than without it. Friction has added to the work needed to slide the weight up the ramp. However, it is often a case of making completion of a task possible. For example, I may not be able to lift a heavy box straight up from the ground to a platform. I could, however, slide it up the ramp. I can overlook the fact that the total work was increased because of friction. The ramp made completion of the task possible.

In many real-world applications, there is a given height to which an object must be raised, and the length of the ramp will determine the angle. This activity will give students the experience necessary to make meaningful choices when faced with a situation where they can control one or more of the variables.

Management
1. There are two parts to this activity. The first is an exploratory exercise in which students will predict and get the feel of using a ramp. The second part will develop that information in more precise terms. The first part takes about 45-60 minutes (including building the model). The second part takes about 60 minutes.

2. Stress that careful attention to the illustrated constructions is very important. Some groups may need to rebuild their model a few times before the results match the picture.
3. The condition of the spring scales, the mass of the resistance, and the skill of the user will have an impact on the data that are collected. Use a spring scale that registers a maximum of 250 grams (2.5 newtons) to get good results.
4. Increasing the mass may help to make the use of the spring scale less frustrating for younger students. A 35 mm film canister filled with sand used along with the weighted brick gives good results.
5. Students need to read their spring scale in the middle of a slow, steady pull. They should pull the scale parallel to the ramp. It may take some practice to develop this technique.
6. Groups of two are suggested for this activity, but larger groups will work if each member is assigned a job (Recorder, Scale Reader, Slope Manager, Brick Controller).
7. Before doing the activity, copy the *Angle Indicator* onto tagboard or have tagboard available on which students can glue the indicator before cutting it out.
8. If data comparison between groups is desired, the axle positions for the ramp support will have to be agreed on before beginning *Part 2*. The illustration's position might be recorded as *Row 3, Hole 5*, along with the angle.
9. Many spring scales are calibrated in newtons and grams. Newton is the correct unit of measurement for force. Before beginning the activity, the teacher will need to decide whether using newtons or grams is most appropriate for the students. Students often have a better "feel" for grams and deal with them as whole numbers rather than a decimal form as with newtons.
10. Taping the base plate to the table may prevent it from lifting or tipping while students are working with the inclined plane.

Procedure
Part 1
1. Draw a picture on the chalkboard showing a heavy box and a platform or deck. Then ask the class, "What would be the easiest way (least effort) to get the box onto the platform." Students are generally familiar enough with ramps that someone will suggest putting boards up to the top of the platform and sliding the box up the boards. Ask them if they have ever seen this done. [Moving vans have ramps that go from the back of the truck to the street.]
2. Direct the students to construct the ramp support from LEGO® elements. This will give them a device with which to systematically change the position of the ramp. It also standardizes the position of the ramp between groups so data may be compared.

3. Have students tape a ruler, paint stir stick, or a similar item, to the top side of the adjustable LEGO® ramp support. This serves as the ramp. The tape should be located at the bottom of the ramp near the connector pegs so it does not get in the way of the sliding brick.

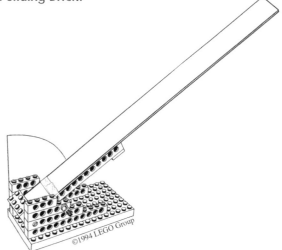

4. Have students attach the *Angle Indicator* as instructed on the indicator page.
5. Direct them to follow the instructions on the first activity sheet and predict and record which ramp positions takes the most effort and the least effort.

Part 2

6. Ask students to discuss their findings from *Part 1*. Pose the *Key Question*.
7. Have the students adjust the ramp so it is at the first position and record the angle of the ramp (in degrees) in the appropriate column on the second activity page. Inform them that the angle is read at the position of the center of the holes on the beam supporting the ruler.
8. Tell the students that they will use the spring scale to find the force generated by the object they are dragging up the ramp. Describe the procedure for slowly and steadily pulling the spring scale parallel to the ramp.
9. Have students predict and record the amount of force they think will be used to pull the resistance at this angle.
10. Have them use a spring scale to measure the amount of force needed to lift the weight straight up. Direct them to record the measurement under *Force of Resistance* on the record chart. (The resistance is the same for all the positions.)
11. Have students use the spring scale to measure the force needed to pull the weight up the ramp. Direct them to record the reading under *Force of Effort*. Have them pull the weight up each ramp for three different trials and record the force each time.
12. Direct them to predict, measure, and record data at different positions to complete the chart.

13. Have them graph the effort force for each position as instructed on the graph page.
14. Allow time for students to calculate the *Mechanical Advantage*.
15. Urge them to analyze and discuss their findings.

Discussion

1. How do the mechanical advantage and the effort compare? [less force, greater mechanical advantage]
2. What does the mechanical advantage tell you about the effort needed to pull a weight up a ramp and the effort just to lift the weight straight up? [how many times the effort force has been multiplied]
3. Describe what patterns you see in the graph? [forms a curve, curve gets steeper as angle gets greater]
4. How is your graph similar to and different from the graphs of other groups? [all should get higher as angles increase, may start at different heights and have different curves]
5. What happens to the force needed to move the resistance as the ramp gets steeper? [gets greater]
6. How does the graph show you that the steeper the ramp, the more effort force it takes to move an object? [It gets higher and steeper.]
7. How do the results on the graph and the mechanical advantage compare? [as force is greater, mechanical advantage is less]
8. How could you use the graph to predict how much effort it would take to move the resistance up a ramp at an angle you have not tried? [connect points and extend curve, interpolate and extrapolate]
9. What is the best angle for the ramp? (Answers will vary, but students must justify their position.)
10. If you had to lift a 200 kilogram crate two meters high, describe the type of inclined plane you would use.

Extensions

1. Put different surfaces (sand paper, plastic, paper) on the inclined plane and see how they affect the force required to move the resistance.
2. Have students find pictures or examples of inclined planes and bring them to class.
3. Investigate the effect of slope on longer inclined planes.

FORCE-UPS

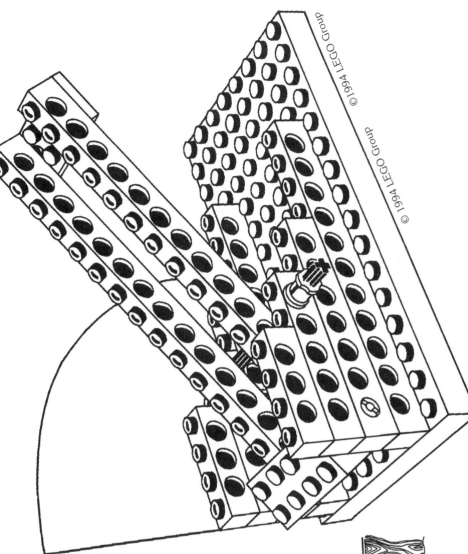

Parts Inventory

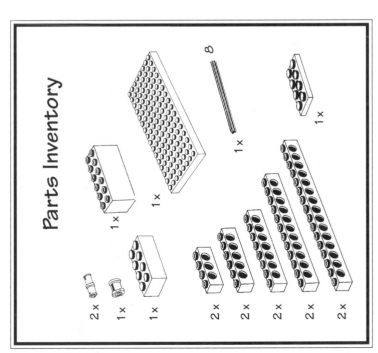

2x
1x
1x

1x

1x

8
1x

1x

2x
2x
2x
2x
2x

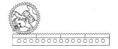

FORCE-UPS

Cut out the *Angle Indicator* and tape it to the model so that the hole for the pivot is lined up with the hole in the *Angle Indicator*.

The ramp is held in position by a movable axle. The axle is placed through the holes in the beams of the model. The ramp support rests on the axle and keeps the ramp in the same position. The angle of the ramp is altered by changing where the axle is placed. The angle of the ramp is measured at the center of the holes where the beam extends past the *Angle Indicator*.

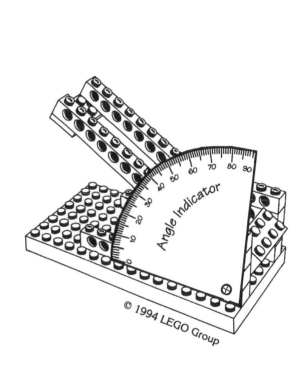

© 1994 LEGO Group

FORCE-UPS

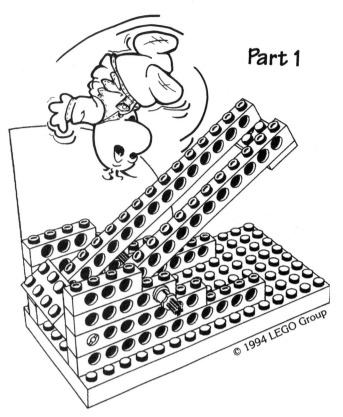

© 1994 LEGO Group

Which of these ramps will make it easiest to pull an object up it? How do you know?

1. Assemble the ramp as shown in the illustration.
2. On the chart below order your prediction of which ramp will require the least to the greatest force to drag a weight to the top. (1 = Least Force, 3 = Greatest Force)
3. Set the ramp in its lowest position. Use your fingers to drag the weight up the ramp. Do the same for the other two positions.
4. Order the actual results on the chart to indicate which positions required the least to greatest force to pull up the weight. (1 = Least Force, 3 = Greatest Force)

Ramp Position	Force to Move Up Ramp	
	Predict	Actual
Steepest		
Steeper		
Steep		

Summarize what you know about the steepness of a ramp and the force used to pull something up the ramp.

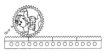

FORCE-UPS

Part 2

How does the force needed to pull a weight up a ramp change as the slope of the ramp changes?

1. Hang your weight from the spring scale and record the measurement under *Force of Resistance*.
2. Place an axle through one of the holes in the beams of the model and rest the ramp support on the axle. Record the *Degrees of Slope* for the ramp.
3. Record your prediction of how much force it will take to pull the weight up the ramp in this position.
4. Put the weight on the ramp. Gently and slowly pull it up the ramp using the spring scale. (Read the force on the scale in the middle of the pull.) Do this for three trials.
5. Calculate the average force and mechanical advantage for this position.
6. Follow steps two to five for other positions.

Position	Degrees of Slope	Force of Resistance	Predicted Effort	Force of Effort				Mechanical Advantage (Resistance ÷ Effort)
				Trial 1	Trial 2	Trial 3	Average	
A								
B								
C								
D								
E								
F								
G								
H								

BRICK LAYERS

7

FORCE-UPS

1. Determine a scale for the vertical axis of the graph that will let you graph the lowest and highest force required to pull the weight up the ramp. Record the scale.

2. Use an **X** to plot each trial at a position and a **dot** to record the average.

3. Connect the dots to form a broken line.

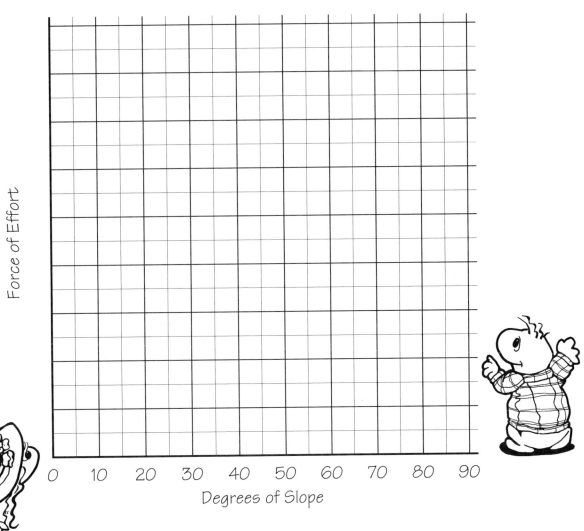

Force of Effort

0   10   20   30   40   50   60   70   80   90

Degrees of Slope

Describe what patterns you find in the graph.

Levers

A lever is a rigid bar which can turn (pivot) around a fixed point called the fulcrum. All levers have a resistance arm, effort arm, and fulcrum. The resistance (load) is the object that needs to be moved and the resistance arm is the part of the lever between the fulcrum and the resistance. The effort is the force that is used in attempting to move the resistance. The effort arm is the part of the lever between the fulcrum and the place where the effort is applied.

Levers can be divided into three classes. It is the arrangement of the three parts of a lever that determines its class.

First-Class Levers

In a first-class lever, the fulcrum is between the effort and resistance. A crowbar is an example of a first-class lever. In the illustration, the fulcrum is the edge of the box used as a pivot point. The resistance is the box lid. The resistance arm is the part of the crowbar under the lid. In the illustration, the boy is applying the effort on the part of the bar that extends past the box. The bar between the boy and the edge of the box is the effort arm.

Resistance   Fulcrum   Effort

As the boy pushes down on the crowbar, the box lid goes up. With a first-class lever the resistance always moves in the opposite direction of the effort.

In this illustration, the effort arm (the distance from the edge of the box to where the boy is pushing) is about twice as long as the resistance arm (the distance from the tip of the crowbar under the lid to the edge of the box). Because the effort arm is twice as long as the resistance arm, the effort's force will be doubled. If the boy pushes down with all his weight, the box's lid will be pushed up with a force that is twice the weight of the boy.

If the boy moved to the very end of the crowbar, he would be four times the distance of the resistance from the fulcrum. Now if he pushed down with all his weight, the resistance would be lifted with a force that is four times his weight. However, if the boy were to move closer to the box, the lever would have the opposite effect. If he pushed down at a distance from the edge that was half the length of the resistance arm, he would lift the resistance with only the force of half his weight.

Depending on where the force of the effort is located, a first-class lever can be an advantage or disadvantage.

Second-Class Levers

The resistance is located between the fulcrum and the effort in a second-class lever. A wheelbarrow is an example of a second-class lever. In this illustration, you can see that the wheel is the fulcrum; the child in the barrow is the resistance; and the girl provides the effort. The resistance arm is the distance from the wheel to the child. The effort arm is the distance from the wheel to the girl.

Fulcrum   Resistance   Effort

When the girl lifts the wheelbarrow, she also lifts the child. In a second-class lever, the effort and the resistance always move in the same direction.

A second-class lever always increases the force of the effort. The girl is farther from the wheel (fulcrum) than the boy. If she is one and a half times further from the wheel than the child, her force will be increased by a factor of one and a half times. This is a real advantage. Assume that the girl can lift twenty pounds but the child weighs thirty pounds. On her own, she can't lift him. When he gets into the barrow, her force is multiplied by one and a half times; her lifting force goes from twenty pounds to thirty pounds. The wheelbarrow helps her do what she could not do by herself.

Third-Class Levers

In a third-class lever, the effort is between the resistance and the fulcrum. Your elbow and lower arm form a third-class lever. The elbow is the fulcrum. Your biceps provide the effort. Whatever you have in your hand is the resistance. If you flex your biceps, you can feel where the muscle attaches with a tendon to your forearm (ulna). The distance from your elbow to the attachment of the biceps is the effort arm. The distance from the elbow to your hand is the resistance arm.

Resistance   Effort   Fulcrum

Your flexing biceps lift your forearm, lifting your hand. In a third-class lever the effort moves the resistance in the same direction it moves.

A third-class lever always decreases the effort's force. The effort applied at the muscle's connection is closer to the elbow than the hand's resistance. If your muscle's connection is one-fourth of your hand's distance from the elbow, your muscle will be able to lift a weight of only one-fourth its strength. If your muscle can lift one hundred pounds, it can only lift a twenty-five pound weight held in your hand. However, if the effort moves an inch, the resistance moves four times as far, four inches.

The third-class lever always is a disadvantage to the effort. Why was your body designed to use a third-class lever? When muscles contract they only shorten a small length with a relatively strong force. To be of much use, your hand has to move a long way. The design of the arm was a logical trade-off. The force made by the strong muscles are reduced to increase the distance the resistance is moved.

Mechanical Advantage

All levers change the force that is applied to them (effort) in the same way, but not by the same amount. The amount the effort force changes can be determined by comparing the effort arm's length to the resistance arm's length. If the effort arm is three times as long as the resistance arm, the effort's force will be increased by a factor of three. If the effort arm is half the length of the resistance arm, its force will be half as strong. The ratio of effort arm's length to the resistance arm's length tells us how much the lever will multiply the effort force. This ratio is called the *mechanical advantage*. If the ratio equals one, the machine does not change the effort force. (Any number times one is the same number.) If the mechanical advantage is greater than one, the lever is helping the effort by increasing its force. When the

mechanical advantage is less than one, the lever is reducing the efforts. In this case it might have been better to call it a "mechanical disadvantage."

The Trade-off

The good news about levers is that they can increase our efforts. The bad news is that you can't get something for nothing. Take a look at the boy who is working hard to pry open the box. The lever is a great help because it is taking the boy's strength and doubling it to open the box. When the boy is done, he has pushed the crowbar a long way, but the box has only been opened a little bit. If you look closely, you will see the boy has moved his side of the bar twice as far as the other side of the bar which moved the box up.

FORCE x distance   =   force x DISTANCE

The work of this simple machine can be thought of as a mathematical equation. The boy is doing his work by taking his small force and pushing the bar a large distance. The crowbar takes the boy's work and changes it into its work with a large force and small distance.

The lever makes a trade of what you have to get what you need. The biceps muscle has a lot of strength but can't go very far so it uses the third-class lever, the arm, as a trade-off. The large force is converted to a smaller force. To do this, the arm trades the small distance the muscle moves to get the hand to move a large distance.

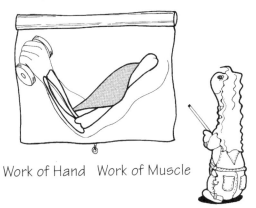

Work of Hand   Work of Muscle

force x DISTANCE   =   FORCE x distance

The lever is one type of simple machine. Like all simple machines, it does not really make work easier. It just changes how the work gets done. It makes an exchange, trading force for distance.

M.V.P.
(Most Valuable Place)

Topic
Simple Machines: Levers

Key Question
How does changing the placement of the center element of a lever affect the operation of the lever?

Focus
Students will explore the workings of a lever. By modifying the structure of the model, they will discover the three classes of levers and arrive at a theory of operation for each.

Guiding Documents
NCTM Standards
- *Describe and represent relationships with tables, graphs, and rules*
- *Reflect on and clarify their own thinking about mathematical ideas and situations*

Project 2061 Benchmarks
- *Sometimes people invent a general rule to explain how something works by summarizing observations. But people tend to over generalize, imagining general rules on the basis of only a few observations.*
- *Organize information in simple tables and graphs and identify relationships they reveal.*

Science
Physical science
    simple machines
        levers
        mechanical advantage

Integrated Processes
Observing
Collecting and recording data
Interpreting data
Generalizing

Materials
LEGO® elements (per group):
  1 base plate
  1 axle, 4-studs long
  6 1 x 2 beams
  1 1 x 16 beam
  1 weighted brick
1 paper clip

Background Information
    A lever is a rigid bar which can turn (pivot) around a fixed point called the *fulcrum*. All levers have a resis-tance arm, effort arm, and fulcrum. The *resistance* is the object that needs to be moved. The *effort* is the force that is used in attempting to move the resistance.

    There are three classes of levers. It is the arrange-ment of the three parts of a lever that determines the class of a lever.

First-Class Lever
- The fulcrum is between the effort and resistance
- Changes the direction of the force and the amount of a force
- Most effective when the fulcrum is closer to the resistance
- Teeter-totter and crowbar are examples

Resistance   Fulcrum   Effort

Second-Class Lever
- The resistance is between the fulcrum and the effort
- Changes only the amount of force
- The resistance is always closer to the fulcrum than the effort
- Most effective as resistance gets closer to the fulcrum
- Wheelbarrow and nutcracker are examples

Fulcrum   Resistance Effort

Third-Class Lever
- The effort is between the resistance and the fulcrum
- Changes only the amount of a force
- The resistance arm is always longer than the effort arm
- Most effective when the effort is closer to the resistance
- Golf club, broom, and arm are examples

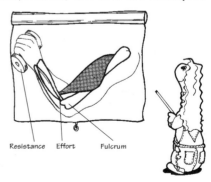

Resistance   Effort   Fulcrum

It is the purpose of these activities to allow the students to explore various positions of the center element of each of the three types of levers, the effect this has on the operation of the lever, and then generalize their findings by writing a rule. The variable throughout the activity is the middle element.

Management

1. This activity is divided into two parts which can be done separately or as one lesson. The first part takes about 45 minutes. In this part, students construct the lever and explore different arrangements of the elements. The second part also takes about 45 minutes for students to observe the effects of moving the middle element in all three types of levers and to make a generalization.

2. Discussions are an important part of this activity. They will allow the teacher to assess the prior knowledge of the class and the vocabulary level at which the class is comfortable. Some groups will already know words like *resistance, effort, fulcrum*. Other groups will use very generic terms to describe these.

3. Writing generalizations is another important part of this activity. Students will generalize their observations with a rule. Those having a vested interest in the correctness of their rules will be very attentive as others read their rules. While nearly all of the sentences will express the same idea, the way in which the ideas are expressed will vary considerably. As students analyze each other's sentences, they are reviewing and reprocessing the information.

4. The teacher may choose to identify the position of the middle element before beginning the activity to get consistent responses from the students. These positions can be marked on the activity pages before copying them for students.

Procedure

Part 1

1. Have students look at the picture of the model that they are about to build. Ask them for any personal experiences they have had using levers. Draw out as much prior knowledge as possible, without comment as to its accuracy.

2. Have students build the first model as shown in the illustration.

3. Have students discuss their initial impressions of the working of the model. Ask them if these match their earlier ideas about how it would work.

4. Identify the three parts of the lever: a. the place to push or pull (effort); b. the pivot point (fulcrum); c. the place where the thing being moved is located (resistance).

5. Distribute the first activity page, and show students that these parts have been labeled on the first recording place.

6. Let students discover what other possible arrangements there might be for the three elements of the lever. The center element will change. (A lever with the pivot on the left end and the resistance on the right end would be considered the same as a lever with the pivot on the right end and the resistance on the left end.)

7. Record the two arrangements on the remaining space provided at the bottom of the page. (One will show the resistance in the center, the other will show the effort in the center.)

Part 2

8. Have the students consider the original model and predict what will happen when the location of the fulcrum is changed. Ask students to discuss their reasoning for their predictions and record them.

9. Have the students do three tests, changing the location of the fulcrum between tests. Direct them to record the exact location of the fulcrum on the record sheet. Guide them to test the effort required to lift the weight with their finger.

10. When students have made all three tests, have them rank and record the levers by the effort required (from hardest to easiest).

11. Allow time for them to analyze the results and generalize their findings by writing a rule about the behavior of levers based on this experience. The language and vocabulary requirements of this rule will depend upon the needs and skills of the class. They should generalize that the closer the fulcrum is to the resistance, the less effort is needed to move the lever.

12. For the other two possible arrangements, have the students reconfigure the model and test them as before. For these two arrangements, students should pull rather than push down. To lift, have students use a bent paper clip with one end through the hole in the beam rather than lifting the beam directly with their fingers.

13. After testing each arrangement, direct the students to write a generalizing rule about the lever. They should find that when the resistance is in the middle, the closer it is located to the fulcrum, the easier it is to pull up. (The closer the load is to the front of a wheelbarrow the easier it is to lift.) They will find that when the effort is in the middle, the closer it is located to the resistance, the easier it is to pull.

Discussion

1. How many ways can the center element (effort, fulcrum, resistance) of a lever be arranged? [3]

2. To make it easiest to lift the weight, where should the fulcrum be placed if it is between the resistance and the effort? [near the resistance]

3. To make it easiest to lift the weight, where should the resistance be placed if it is between the effort and the fulcrum? [near the fulcrum]

4. To make it easiest to lift the weight, where should the effort be placed if it is between the resistance and the fulcrum? [near the resistance]

5. What are some real-world examples of the three types of levers?

6. Which types have you used today?

Extension

Students may bring in pictures, draw pictures, or make lists of levers that they find around the home or in the classroom. An area can be set aside for this display to grow as the study of levers progresses.

M.V.P.
(Most Valuable Place)

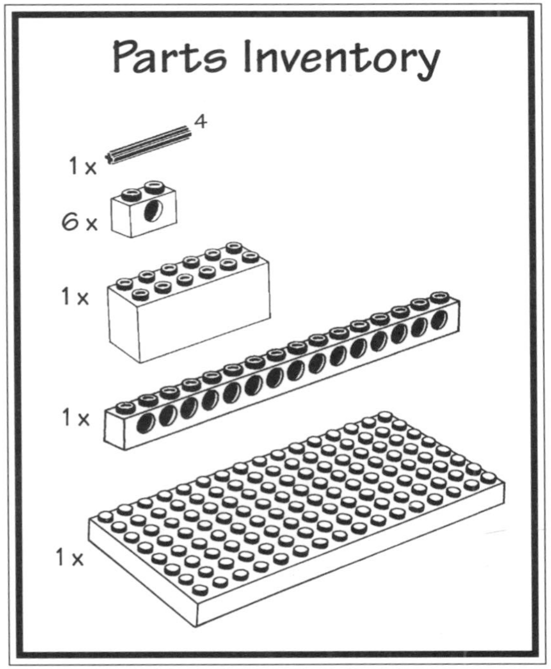

Parts Inventory
1 x
4
6 x
1 x
1 x
1 x

2

© 1994 LEGO Group

1

© 1994 LEGO Group

3

© 1994 LEGO Group

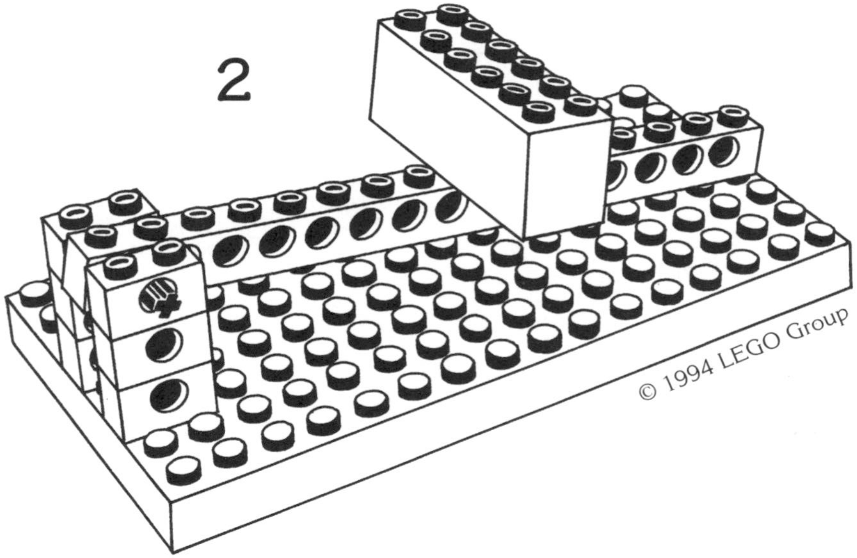

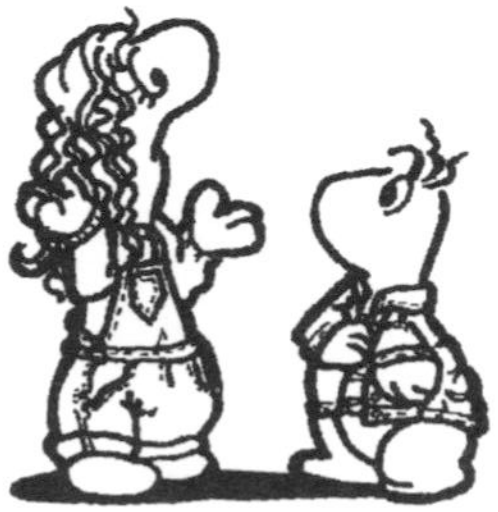

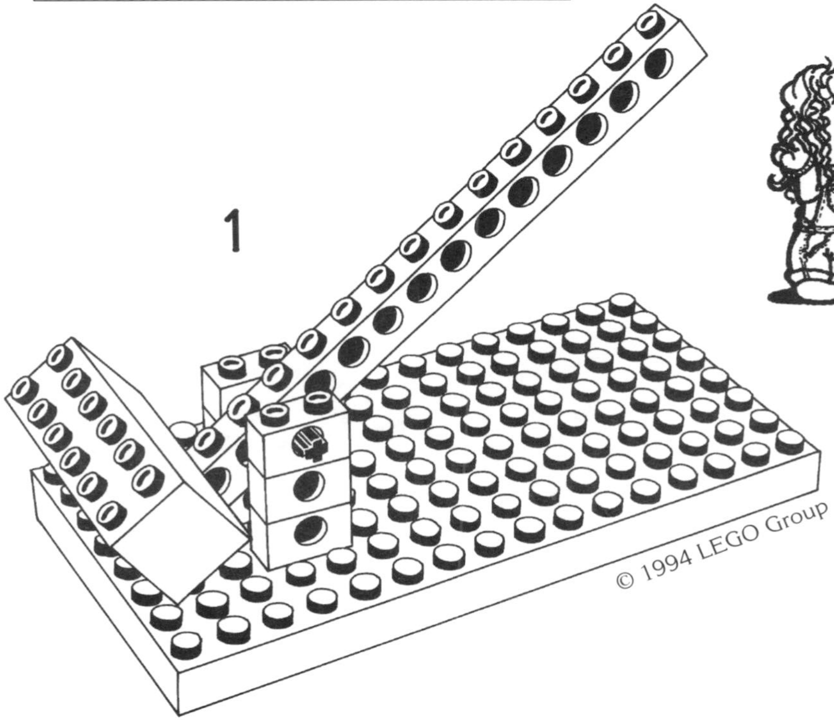

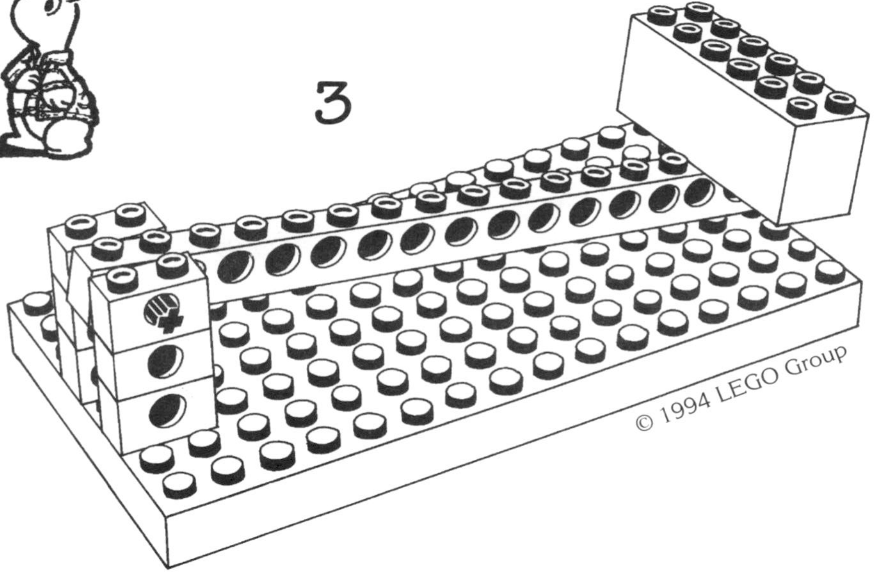

M.V.P.
(Most Valuable Place)

A LEVER has three important places.
The EFFORT is the place where you push or pull the lever.
The FULCRUM is the place around which the lever turns.
The RESISTANCE is the place where the object being moved is located.

The three important positions on the lever in the drawing are labeled on the first beam below. There are two other possible ways to arrange the positions on the beam of a lever. Determine the other arrangements and label them on the two remaining beams.

Resistance	Fulcrum	Effort

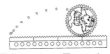

M.V.P.
(Most Valuable Place)

If the fulcrum is between the resistance and the effort, where should it be placed to make it easiest to lift the weight?

© 1994 LEGO Group

Make three tests, moving the position of the fulcrum each time. Use an **X** to mark the hole where you put the fulcrum on each trial. Push down on the lever to determine the positions that make it easiest and hardest to lift the resistance. Rank the positions: 1 = easiest to 3 = hardest.

Rank

Test 1:

Resistance   Effort

_____

Test 2:

Resistance   Effort

_____

Test 3:

Resistance   Effort

_____

Summarize what you have discovered:

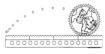

M.V.P.
(Most Valuable Place)

If the resistance is between the fulcrum and the effort, where should it be placed to make it easiest to lift the weight?

© 1994 LEGO Group

Make three tests, moving the position of the weighted brick each time. Use an **X** to mark the location of the weighted brick for each trial. Pull up with a paper clip to determine the positions of the brick that make it easiest and hardest to lift. Rank the positions: 1 = easiest to 3 = hardest.

Rank

Test 1:

Fulcrum

Effort

_____

Test 2:

Fulcrum

Effort

_____

Test 3:

Fulcrum

Effort

_____

Summarize what you have discovered:

M.V.P.
(Most Valuable Place)

If the effort is between the resistance and the fulcrum, where should the effort be placed to make it easiest to lift the weight?

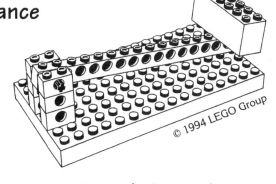

© 1994 LEGO Group

Make three tests, each time moving the position where the effort is applied. Use an **X** to mark the hole where you applied the effort. Pull up with a paper clip to determine, which positions make it easiest and hardest to lift the resistance. Rank the positions: 1 = easiest to 3= hardest.

Test 1:

Fulcrum   Resistance

Rank

_____

Test 2:

Fulcrum   Resistance

_____

Test 3:

Fulcrum   Resistance

_____

Summarize what you have discovered:

Fiddling with Fulcrums

Topic
Simple Machines: First-Class Levers

Key Question
Where should you place the fulcrum to make it easiest to lift a heavy object?

Focus
Students will study how the position of the fulcrum affects the amount of effort needed to lift a consistent resistance.

Guiding Documents
NCTM Standards
- *Represent numerical relationships in one- and two-dimensional graphs*
- *Analyze tables and graphs to identify properties and relationships*
- *Analyze functional relationships to explain how a change in one quantity results in a change in another*

Project 2061 Benchmarks
- *If more than one variable changes at the same time in an experiment, the outcome of the experiment may not be clearly attributable to any of the variables. It may not always be possible to prevent outside variables from influencing the outcome of an investigation (or even to identify all of the variables), but collaboration among investigators can often lead to research designs that are able to deal with such situations.*
- *Graphs can show a variety of possible relationships between two variables. As one variable increases uniformly, the other may do one of the following: always keep the same proportion to the first, increase or decrease steadily, increase or decrease faster and faster, get closer and closer to some limiting value, reach some intermediate maximum or minimum, alternately increase and decrease indefinitely, increase and decrease in steps, or do something different from any of these.*

Math
Using computation
Estimating
Graphing
Averaging
Using formulas

Science
Physical science
   simple machines
      first-class lever
      effort
      resistance

Integrated Processes
Observing
Collecting and recording data
Interpreting data
Generalizing

Materials
LEGO® elements (per group):
  1 axle, 4-studs long
  6 1 x 2 beams
  1 base plate
  1 weighted brick
  1 1 x 16 beam
String
1 paper cup
100 centicubes
Balance

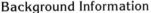

Background Information
In this activity students will develop their knowledge of a first-class lever. In a first-class lever, the fulcrum is between the *effort* and the *resistance*. The longer the effort arm is compared to the resistance arm, the greater the *mechanical advantage*. While the mechanical advantage may be greater, the *distance* through which the effort travels is also increased. The formula W = F x D (Work = Force x Distance) applies. The greater the distance through which the effort is exerted, the less effort is needed to accomplish the same amount of work.

The mechanical advantage is the answer to "How much is this machine helping me?" It is the relationship between the force of the resistance (weight of the load) and the force of the effort. (MA=Force of resistance÷Force of effort). The greater the mechanical advantage, the easier the machine is to operate.

In this activity, students will come to recognize that when a lever is equal length on both sides of the fulcrum, there is no mechanical advantage. As the effort side of the lever gets longer compared to the resistance side, it takes less and less effort to move the resistance. Conversely as the resistance side gets longer than the effort side, it takes even more force to move the lever.

Management
1. Allow one and a half hours for students to build the model, make preliminary measurements, gather data, share data, analyze, and discuss findings.
2. Groups of two students are recommended for this activity.

3. Recognizing and controlling variables are important parts of experimentation. Discuss the variables involved in this activity and encourage students to be very careful in controlling them so their data are consistent. Some of the complications that may result in inaccurate data are: improperly built models, inaccurate counting of cubes, the lever gets stuck on the base plate, the cup hangs against the edge of the table, use of too much force when putting cubes into the cup.

4. Tell students that they should add mass to the lever until it tips, not just moves. Warn them to keep a hand on the model to prevent it from spilling onto the floor.

5. To ease counting of centicubes, students may want to put them together in trains of ten.

6. One hundred centicubes is generally enough to complete the activity. If a small, lightweight cup is used, it may take more. Be prepared for this with extra centicubes or have students share materials.

7. Make a transparency or large chart of *Fiddling with Fulcrums-Group Data, Class Data* to use for recording group results.

8. Decide whether students are to graph the class average data in bar or broken-line format.

Procedure

Construction

1. Discuss the *Key Question* with students. Have them list variables that will affect the answer to the question such as length of lever arms and placement of fulcrum.

2. Have students build the model of the lever shown in the illustration.

3. Guide them to tie and hang the cup from the beam as shown in the illustration.

Part 1

4. Have students predict the number of centicubes it will take to tip the lever with the fulcrum in three different positions (five, seven, and nine are suggested) and record their predictions on the group data chart.

5. Direct them to place the fulcrum in each of these positions and add centicubes until the lever tips. Have them record the number of centicubes added on the group data chart.

6. Have groups compare their data from the points and record them on the class data chart. Their investigations should have produced nearly identical results. If there is a group that is lacking in technique or is misunderstanding the directions, then this will bring it to their attention. (Discrepancies in data provide an excellent opportunity for discussion about controlling variables. Questionable data should be retested. Taking care of this in the beginning, when only three positions are being tested, is good because there is ample data to confirm proper procedure, and data gathered later that will be used for graphing will be more accurate.)

7. Have students make predictions for the remaining positions and then test to see how many cubes it takes to move the lever.

8. Have the groups share the data and record it on the class data chart.

9. Direct students to find the average of the number of cubes for each position and record it.

10. Have them graph the average number of cubes for each position. Tell students which type of graph to use, bar graph or a broken-line graph.

Part 2

11. Have the students follow the directions on the student page and complete the chart to determine mechanical advantage.

12. Direct them to write a summary of what pattern they found in the number of cubes it took to move the lever when the fulcrum was in different positions.

Discussion

1. Where was the fulcrum placed when the lever was easiest to move? [closest to the weighted end]

2. Where was the fulcrum placed when the lever was hardest to move?[farthest from the weighted end]

3. What happened as the weighted end got farther from the fulcrum?[[harder to lift]

4. What happens to the graphed line as the position of the resistance gets a greater distance from the fulcrum? [The graphed line (or bars) gets higher.]

5. Each time you move the fulcrum one position, do you add the same amount of cubes to the cup? Explain.

6. Is there a pattern to the increase of cubes needed for each hole further from the fulcrum? Explain. [Theoretically there is, but with data gathered it may not be discernible.]

7. From your findings in this investigation, what causes the mechanical advantage to increase? [longer effort arm in relationship to resistance arm]

8. Explain in your own words the meaning of mechanical advantage.

Extensions

1. Have students use their data and graph to predict what will happen when the fulcrum is placed at the second, third, eleventh, and twelfth positions. Have students justify and test their predictions.

2. Have students gather pictures or make a list of different levers they can identify.

Fiddling with Fulcrums

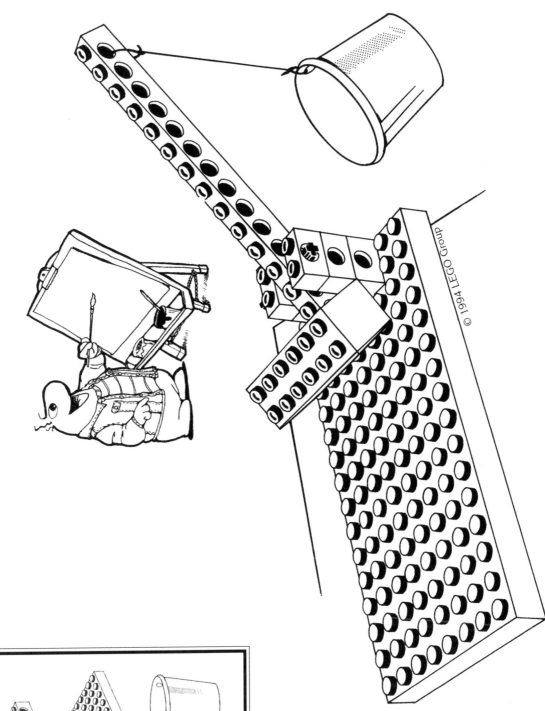

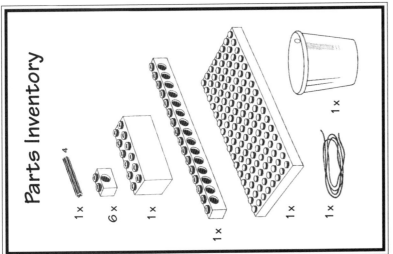

Parts Inventory

1 x

6 x

1 x

1 x

1 x

1 x

Fiddling with Fulcrums

Predict how many cubes it will take to move the lever at each position and record it on the chart. Put the fulcrum at each position and gently put cubes into the cup until it tips the weight. Record the actual number of cubes it took. Share your group's data with the class and find a class average for each position. Make a graph using the class average.

Group Data

	4	5	6	7	8	9	10
Position: Numbers of holes from weighted end to fulcrum							
Prediction: How many cubes will it take to move the lever?							
Actual: How many cubes did it take to move the lever?							

Class Data

	Position						
	4	5	6	7	8	9	10
Group 1							
Group 2							
Group 3							
Group 4							
Group 5							
Group 6							
Group 7							
Group 8							
Group 9							
Group 10							
Total							
Average							

Fiddling with Fulcrums

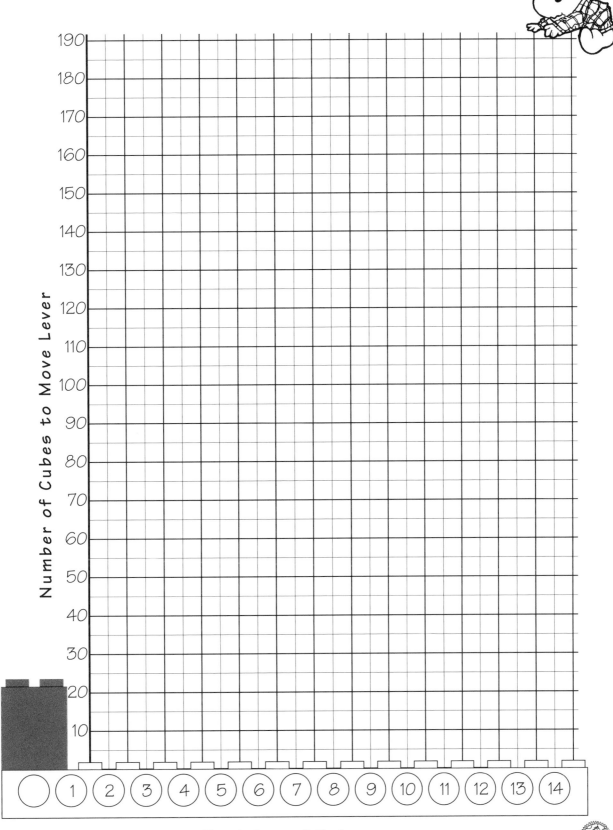

Number of Cubes to Move Lever

190
180
170
160
150
140
130
120
110
100
90
80
70
60
50
40
30
20
10

( ) 1 2 3 4 5 6 7 8 9 10 11 12 13 14

Position of Fulcrum

Fiddling with Fulcrums

Where should you place the fulcrum to make it easiest to lift a heavy object?

Using a balance, find out how many cubes along with the cup and string it takes to equal the mass of the weighted brick. Record the number of cubes on the chart in the column labeled *Mass of the brick in cubes*. Complete the rest of the chart to calculate the *Mechanical Advantage*.

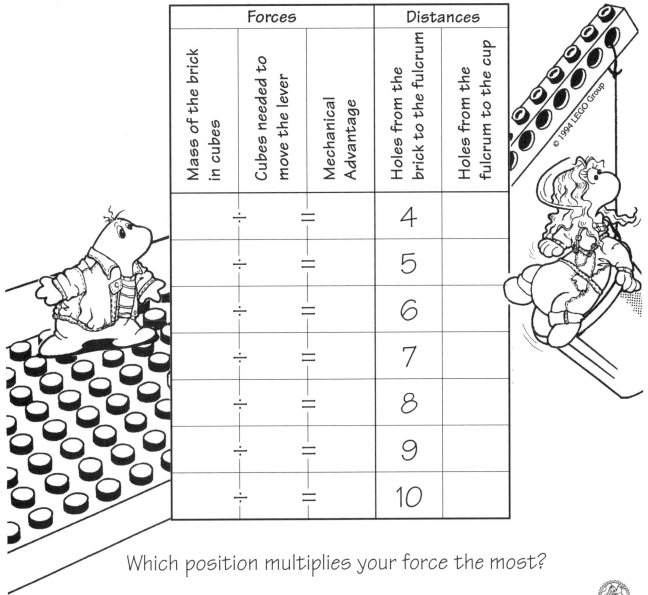

Forces			Distances	
Mass of the brick in cubes	Cubes needed to move the lever	Mechanical Advantage	Holes from the brick to the fulcrum	Holes from the fulcrum to the cup
÷	=		4	
÷	=		5	
÷	=		6	
÷	=		7	
÷	=		8	
÷	=		9	
÷	=		10	

Which position multiplies your force the most?

© 1994 LEGO Group

Beams Over Board

Topic
Simple Machines: First-Class Levers

Key Question
How can you determine where to put a mass to balance a lever before setting the mass on the lever?

Focus
Students will discover the law of the lever by determining where to place different masses on a lever to balance it.

Guiding Documents
NCTM Standards
- *Represent situations and number patterns with tables, graphs, verbal rules, and equations and explore the interrelationships of these representations*
- *Analyze tables and graphs to identify properties and relationships*

Project 2061 Benchmarks
- *Mathematical statements can be used to describe how one quantity changes when another changes. Rates of change can be computed from magnitudes and vice versa.*
- *Sometimes people invent a general rule to explain how something works by summarizing observations. But people tend to overgeneralize, imagining general rules on the basis of only a few observations.*

Math
Using computation
Estimating
Using and applying formulae

Science
Physical science
    simple machines
        first-class lever
        law of the lever

Integrated Processes
Observing
Collecting and recording data
Interpreting data
Generalizing

Materials
LEGO® elements (per group):

1	axle, 6-studs long
2	bushings
12	1 x 2 beams
3	1 x 16 beams
1	weighted brick
2	1 x 6 plates
2	1 x 8 plates
1	2 x 6 plate
1	base plate

Background Information
The *law of the lever* is fundamental to this activity. It states that a lever will be balanced when the product of the force applied to one side of the lever and the distance that force is from the fulcrum is equal to the force and distance applied to the other side of the lever. In this activity, *force* is measured in number of LEGO® beams and *distance* is measured in number of studs on a LEGO® element.

An example of the mathematics involved in this activity: On the left side, one beam placed 20 holes from the center of the lever needs two beams placed ten holes from the center on the right side in order to balance (1 x 20 = 2 x 10).

Management
1. Allow 60 to 90 minutes for students to build the model, make measurements, analyze the data, and discuss their findings.
2. Even when the bricks are placed properly, the lever may not always balance. Students may find this frustrating. Have them determine the position for the bricks where the lever is closest to balanced.
3. Distance along the balance is measured in hole spaces. Make sure students count the hole space where the beams join.

Procedure
1. Have the students build the lever as shown in the illustration.
2. Discuss the *Key Question*.
3. Direct the students to hang a beam on the left side of the fulcrum at the position specified in the chart.

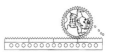

4. Have them find the position on the right side of the lever on which to hang a beam to balance the lever. Direct them to record this position on the chart.

5. Tell the students to follow a similar procedure to complete the chart. When two or more beams are required, they should be stacked one on top of the other so they may be hung at one position.

6. Have students find patterns in their charts. They may use the two empty columns in the middle as helping columns in which to write the product of the number of beams times the number of holes from the fulcrum.

7. Discuss the findings.

Discussion

1. In each section of the chart, what happens as you put more beams on the right side of the balance? [The beams are placed closer to the fulcrum.]

2. What can you do with the two numbers on the right side to equal the numbers on the left side? (If students are having difficulty, suggest they try adding, subtracting, multiplying, and dividing.) [Multiply them.]

3. How can you determine the missing number without using the balance? [Multiply the two numbers on the left. Divide the product by the known number on the right to get the unknown number on the right.]

4. From doing this activity, explain in your own words what you have learned about balancing the levers?

Extension

Have students determine how to hang more than one pile of beams on the right side. (Example: The two beams at the eighteenth hole are balanced by three beams at the fourth hole, and two beams at the twelfth hole.)

Beams
Over Board

1

2

3

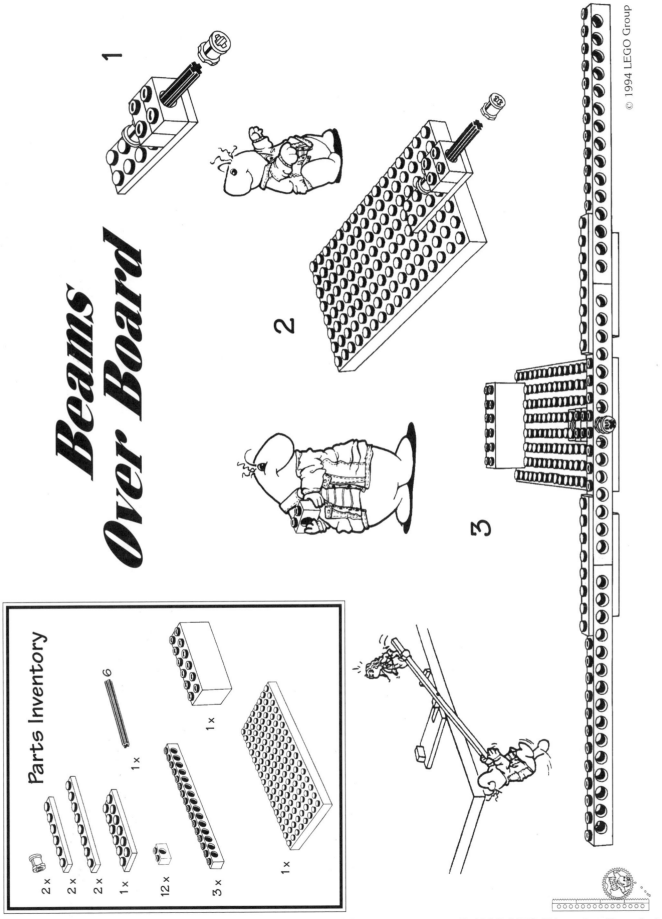

Parts Inventory

2 x
2 x
2 x
1 x
6
1 x
12 x
3 x
1 x
1 x

Beams Over Board

1. Build the balance in the illustration.

2. On the left side of the balance, centered under the hole, hang the number of beams listed in the table.

3. Find the place or number of beams needed to be hung on the right side of the fulcrum to level the balance.

4. Record the missing number on the right side of the chart.

5. Study the chart and summarize below how you could determine where to place the beams on one side to make it balance the other side.

Left Side				Right Side	
Holes from Fulcrum	Beams to Balance			Beams to Balance	Holes from Fulcrum
20	1			1	20
20	1			2	10
20	1			4	5
20	1			5	4
20	2			2	20
20	2			4	10
20	2			5	8
20	2			8	5
18	2			2	18
18	2			3	12
18	2			4	9
18	2			6	6
15	2			2	15
15	2			3	10
15	2			5	6
15	2			6	5
12	2			2	12
12	2			3	8
12	2			4	6
12	2			6	4
12	2			8	3

Effort-Less

Topic
Simple Machines: Levers

Key Question
What happens if you move only the effort position on a lever?

Focus
In this activity, students move the position the effort is applied to a lever to see how the position affects the amount of effort required. They will become familiar with second- and third-class levers, and will be introduced to the concept of mechanical advantage as they graph and analyze their data.

Guiding Documents
NCTM Standards
- *Represent numerical relationships in one- and two-dimensional graphs*
- *Analyze functional relationships to explain how a change in one quantity results in a change in another*
- *Analyze tables and graphs to identify properties and relationships*
- *Estimate, make, and use measurements to describe and compare phenomena*

Project 2061 Benchmarks
- *Energy cannot be created or destroyed, but only changed from one form into another.*
- *Graphs can show a variety of possible relationships between two variables. As one variable increases uniformly, the other may do one of the following: always keep the same proportion to the first, increase or decrease steadily, increase or decrease faster and faster, get closer and closer to some limiting value, reach some intermediate maximum or minimum, alternately increase and decrease indefinitely, increase and decrease in steps, or do something different from any of these.*
- *Sometimes people invent a general rule to explain how something works by summarizing observations. But people tend to over generalize, imagining general rules on the basis of only a few observations.*

Math
Measuring
    force
Graphing
Identifying patterns

Science
Physical science
    simple machines
        2nd-and 3rd-class levers
        mechanical advantage

Integrated Processes
Observing
Predicting
Comparing and contrasting
Collecting and recording data
Interpreting data
Generalizing

Materials
LEGO® elements (per group):
  1 axle, 4-studs long
  3 1 x 2 beams
  4 1 x 4 beams
  2 1 x 16 beams
  2 1 x 6 plates
  1 base plate
  1 weighted brick
1 spring scale (250g or 2.5 newtons)

Background Information
In both the second- and third-class levers, the effort and resistance are on the same side of the fulcrum; they share the lever arm. In the *second-class lever*, the effort is farther away from the fulcrum than the resistance. In the *third-class lever*, the effort is nearer to the fulcrum than the resistance.

In this activity, the resistance is placed in the middle of a lever arm. Students move the effort from the position farthest from the fulcrum to one nearest the fulcrum. As the position the effort is applied changes, the lever changes from a second-class to a third-class lever. At the beginning when the effort is farther from the fulcrum than the resistance, it is a second-class lever. As the effort is moved towards the fulcrum and passes the resistance, it becomes a third-class lever.

As the effort gets closer to the fulcrum, it takes an ever greater force to lift the lever. When lifting the lever at the position directly over the weighted brick, the force is the same as lifting the brick and lever arm.

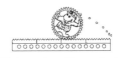

The concepts of changing effort force along a lever arm is easily illustrated with a graph. Those effort positions closer to the fulcrum than the resistance will have bars for the effort force of greater height than the bars for the effort needed to lift directly at the resistance. Effort positions farther from the fulcrum than the resistance will have bars for the effort force shorter than the bars for the effort needed to lift directly at the resistance.

Those positions that are farther from the fulcrum than the weighted brick take less force to lift than the position at the brick. Those positions give an advantage to the person lifting the lever. The person uses less force than the force needed to lift the brick and arm directly. This advantage is called the *mechanical advantage*. It is calculated by comparing the force at the resistance position to the force at the effort position. For example, if it took 0.6 newtons to lift the brick and arm directly, and 0.3 newtons to lift it at the thirtieth position, there would be a mechanical advantage of two ($0.6 \div 0.3 = 2$). This mechanical advantage says that the effort force is doubled by using the lever with these positions.

If the effort position is moved to the fifth hole from the fulcrum, the effort has to increase greatly in this position, say to 1.8 newtons. In this position, the mechanical advantage is one-third ($0.6 \div 1.8 = 0.33...$). In these positions, the lever allows the effort force to lift a resistance of one-third its force.

When using the force measurements to calculate mechanical advantage, one is determining the *actual mechanical advantage*. This is what really happens after friction and other variables have reduced the force. If one wants to predict what will happen before using a lever, the *ideal mechanical advantage* is calculated. This is determined by comparing the effort arm's length to the resistance arm's length. If the spring scale is placed at the thirtieth hole and the brick is at the fifteenth position, the ideal mechanical advantage is two ($30 \div 15 = 2$).

Management
1. Allow one and a half hours for this activity.
2. Using the spring scale in this activity requires some technique and practice. Students should put the scale's hook through the appropriate hole and pull the lever arm up until the arm is level. The reading is taken at this level position.
3. Students will find they have more accuracy if they start at position 30 and work in towards the fulcrum.
4. As students get to the positions near the fulcrum, they will need to hold the base down. At positions one, two, and three, the force is greater than the spring scale registers. These positions do not need to be done.
5. For students to get a measurement at the position

of the resistance (position 15), they need to hook the beam that is on top of the weighted brick.
6. The activity is written using measurement units of newtons. If the spring scale measures only in grams, or it is more appropriate for students to make measurements with whole numbers, have the data recorded in grams and adjust the scales on the graphs. (Although grams are used as a measure of mass, not force, the data will be similar.)
7. An activity sheet that incorporates mechanical advantage is included if these calculations are appropriate for the students. In the *Comparing Force* chart, the actual mechanical advantage is calculated. In the *Comparing Distance* chart, the ideal mechanical advantage is calculated.
8. A second graph is included for use after the initial activity. It provides an excellent assessment. Students cut out a picture of the weighted brick, place it along the picture of the lever arm on the graph, predict what the graph will look like, and explain their reasoning. Then they adjust the lever to check their prediction.

Procedure
1. If *MVP (Most Valuable Place)* has been done, have students review their conclusions about levers. Reinforce the idea that while a scientist may "have a feeling" this is how things work, there is often a need to collect measurable data to give substance to the "feeling."
2. Have students look at the illustration of the lever on the model page and discuss the *Key Question*.
3. Direct them to build the model of the lever shown in the illustration.
4. Show students how to put the spring scale hook through the thirtieth hole from the fulcrum and lift the lever until the arm is parallel with the table.
5. Have students measure the amount of force needed to keep the lever in this position and record the measurement in the corresponding space at the bottom of the graph.
6. Point out to students that this measurement is graphed by coloring the bar above the position to represent the measured force.
7. Have students follow the same procedure for the remaining positions.
8. If appropriate, direct them to transfer the necessary data onto the mechanical advantage charts and calculate the mechanical advantage of different positions.
9. Have students write a summary and discuss the results of the data they have collected and graphed. If they have done *MVP (Most Valuable Place),* they should discuss whether these results agreed or disagreed with their original hypothesis.
10. Students may do the extra graph as an extension or assessment (see *Management).*

Discussion

1. What patterns do you see in the graph? [greater effort with lower position, greater "jumps" with lower positions]

2. A second-class lever has the resistance located between the effort and the fulcrum. On the levers we made, how do you know when you've made a second-class lever? [when the effort position is greater than 15.]

3. How do the forces of positions greater than 15 compare to the force at 15? [They are smaller.]

4. A third-class lever has the effort between the resistance and the fulcrum. On the levers we made, how do you know when you've made a third-class lever? [when the effort position is less than 15]

5. How do the forces of positions less than 15 compare to the force at 15? [They are greater.]

6. Ask questions such as:
   - How many times harder is it to lift the lever at position five than at position 15? [(Answers will vary but it should be about three times harder. Direct counting or comparison may be necessary for students.)]
   - How does the force at position 30 compare to the force at position 15? [about half]

7. What is the advantage of a second-class lever? [You can lift something with less force.]

8. What is the advantage of a third-class lever? [For students, the increase in force needed is a mechanical "disadvantage." The reason for choosing a third-class lever is to increase the speed or distance the lever travels.]

9. Follow a line of questioning such as:
   - How much more force is needed to lift the lever at position 15 than at position 30? [2 times]
   - How much farther is position 30 from the fulcrum than position 15? [2 times]
   - What fraction of force is needed to lift the lever at position 15 compared to position five? [one-third]
   - What fraction of distance is position five from the fulcrum compared to position 15? [one-third]

10. How could you use the positions on the lever to predict how much force will be needed to lift the lever? [Use the reciprocal of the distance relationship: twice the distance, half the force; one-third the distance, three times the force.]

Extension

Have students collect pictures of different levers. Have them identify the class of lever and advantage of each selection.

Effort-Less

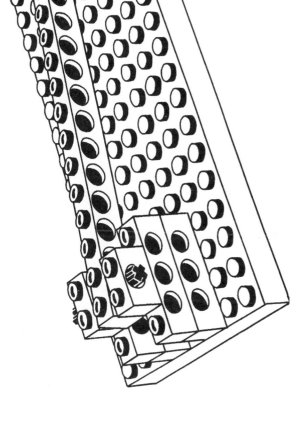

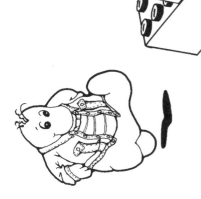

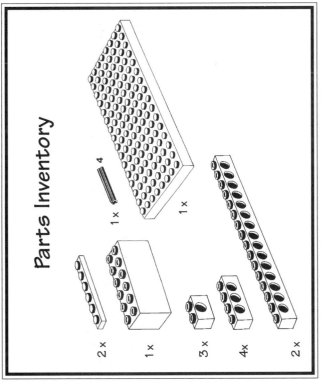

© 1994 LEGO Group

Parts Inventory

2 x

1 x

3 x

4 x

2 x

1 x

1 x

4

Effort-Less

Effort Force (Newtons)

3.0
2.5
2.0
1.5
1.0
0.5

Effort Force at Position (Newtons)

Fulcrum

① ② ③ ④ ⑤ ⑥ ⑦ ⑧ ⑨ ⑩ ⑪ ⑫ ⑬ ⑭ ⑮ ⑯ ⑰ ⑱ ⑲ ⑳ ㉑ ㉒ ㉓ ㉔ ㉕ ㉖ ㉗ ㉘ ㉙ ㉚

Effort-Less

Comparing Distance		
Scale to Fulcrum	÷ **Brick** to Fulcrum	= Ratio **Scale** / **Brick**
30	15	
25	15	
20	15	
15	15	
10	15	
5	15	

Position of Scale

- 30
- 25
- 20
- 15
- 10
- 5
- Fulcrum

Comparing Force		
Force on **Brick**	÷ **Force** at **Scale**	= Ratio **Brick** / **Scale**

How does the position of the effort affect the amount of force required?

© 1994 LEGO Group

Effort-Less

Cut out picture of the weighted brick and glue it to the beam where you plan to put it.

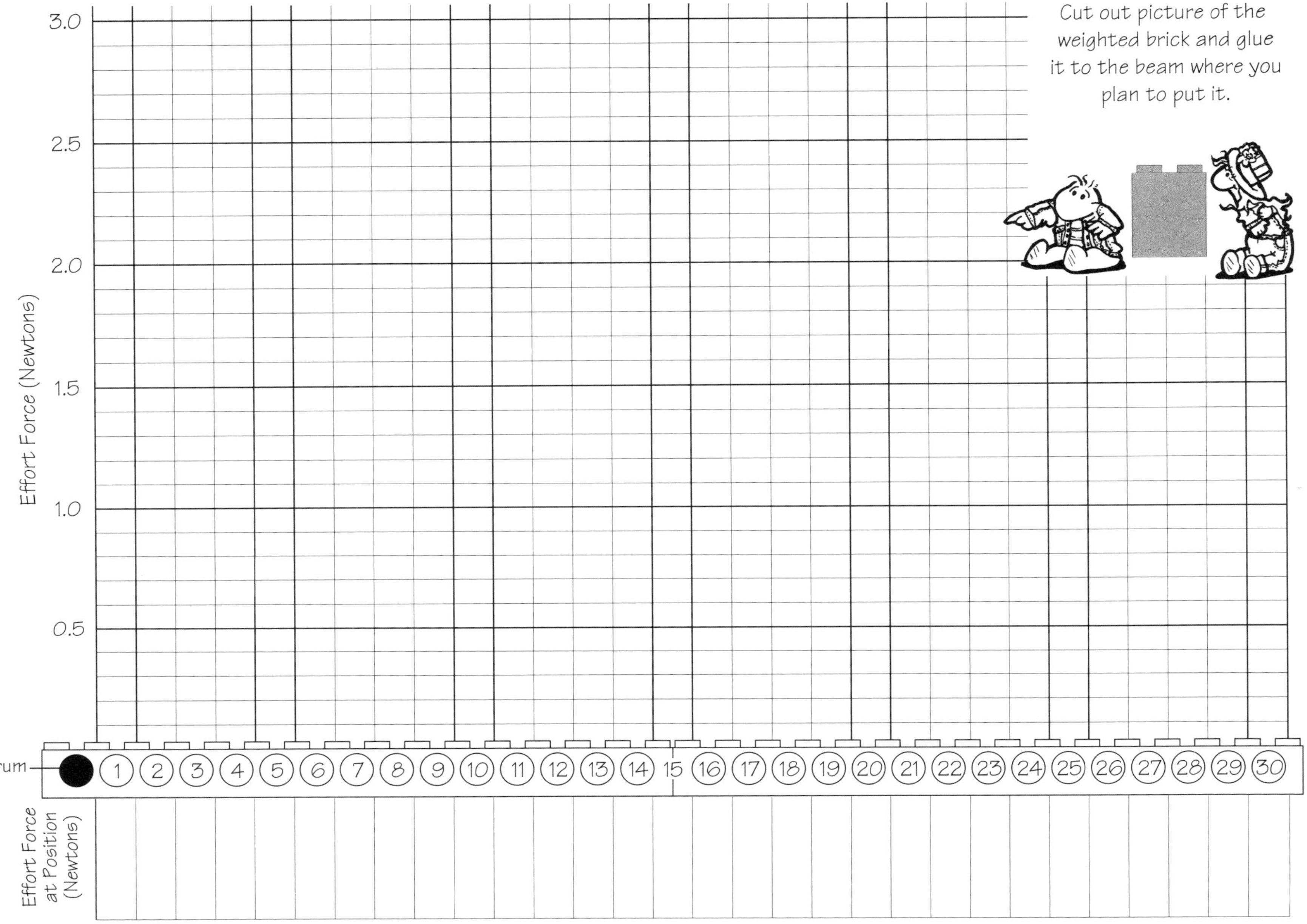

Effort Force (Newtons)

3.0

2.5

2.0

1.5

1.0

0.5

Effort Force at Position (Newtons)

① ② ③ ④ ⑤ ⑥ ⑦ ⑧ ⑨ ⑩ ⑪ ⑫ ⑬ ⑭ ⑮ ⑯ ⑰ ⑱ ⑲ ⑳ ㉑ ㉒ ㉓ ㉔ ㉕ ㉖ ㉗ ㉘ ㉙ ㉚

Fulcrum

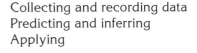

Bear-Barrow and Challenge

Topic
Simple Machines: Levers

Key Question
How will you load the bear-barrow to make it the easiest to move?

Challenge
Design a wheelbarrow type vehicle that requires the least effort to use.

Focus
Students will apply what they have learned in previous investigations to the operation of a second-class lever.

Guiding Documents
NCTM Standards
- *Formulate problems from situations within and outside mathematics*
- *Verify and interpret results with respect to the original problem situation*

Project 2061 Benchmarks
- *Design usually requires taking constraints into account. Some constraints, such as gravity or the properties of the materials to be used, are unavoidable. Other constraints, including economic, political, social, ethical, and aesthetic ones, limit choices.*
- *Inspect, disassemble, and reassemble simple mechanical devices and describe what the various parts are for; estimating what the effect that making a change in one part of a system is likely to have on the system as a whole.*

Math
Measuring
  force
Identifying patterns

Science
Physical science
  simple machines
    levers
    mechanical advantage

Integrated Processes
Observing
Comparing and contrasting
Collecting and recording data
Predicting and inferring
Applying

Materials
LEGO® elements (per group):
  2 1 x 16 beams
  2 1 x 12 beams
  2 1 x 6 beams
  3 1 x 4 beams
  2 1 x 2 beams
  4 connector pegs
  1 axle, 8-studs long
  1 axle, 10-studs long
  2 wheels with tires
  2 1 x 8 plates
  1 2 x 6 plate
  2 bushings
Teddy Bear Counters
Tagboard
Spring scales
Tape
Scissors
One-hole punch

Background Information
A *second-class lever* has the resistance between the fulcrum and the effort. A wheelbarrow is an example of a second-class lever. Second-class levers change only the amount of force. They become more effective (require less force) as the resistance gets closer to the fulcrum.

Management
The following approach is offered for those students ready for more independent exploration.

> *Open-ended:* Distribute *Paul Bunyan's Bear-Barrow Challenge* and have students devise their own plan for designing and testing their solutions. Ask them to record their design and data. Have students display their solutions and discuss their findings.

1. This activity should take about 40 minutes. Students will form hypotheses and use the experimentation to confirm or adjust their hypotheses.

2. Groups of two work well for this activity.
3. Measurements of force should be made in newtons. If students choose to use grams, their conclusions will be the same.
4. Before the activity, copy the seat pattern onto tagboard.
5. If students have difficulty with Teddy Bear Counters sliding around, have them use loops of tape to secure the bears to their positions.
6. As a literature connection, read or have students read stories about Paul Bunyan.

Procedure

1. Distribute the *Paul Bunyan's Bear-Barrow* sheet and have the students read the story.
2. Discuss the *Key Question*.
3. Distribute the materials and have students build the bear-barrow.
4. Direct students to construct the seating pattern by cutting it out, punching holes in both ends, cutting the bold tab lines, folding on the dashed lines, and taping the tabs to form a shallow box. Have them attach the seating pattern to the bear-barrow.
5. Ask students to predict where to place eight bears in the barrow to make it easiest to lift. Direct them to also predict how much force will be required to lift the barrow. Have them record the positions of the bears on the record sheet and then place them in the barrow.
6. Tell the students to use the spring scale to lift the barrow by its handle. Instruct them to record the force needed to lift it.
7. Urge the students to try at least two other arrangements and record the results.
8. Have them summarize their findings.

Discussion

1. When looking at other groups' data, what was similar and different about their arrangements? (There may be more than one arrangement that has the same, or nearly the same results. All results requiring the least effort, however, will have the bears grouped toward the front of the barrow.)
2. What general rule could you make about loading the barrow to best use the mechanical advantage? [Fill it from the front.]
3. When the first two rows are filled, the ninth bear must go in the third row. Does it matter which of the seats in that row it occupies? Explain. [There is no difference in mechanical advantage for any seats in that row.]
4. Why is it important to try more than one position? [Through testing several positions, you can support or reject the hypothesis.]

Extension

Bring in a wheelbarrow and let students feel how the placement of the load affects the force needed to lift the barrow.

Paul Bunyan's
Bear-Barrow and Challenge

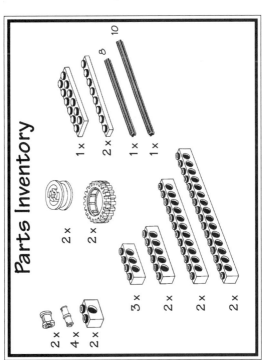

Parts Inventory

2 x
4 x
2 x

3 x
2 x
2 x
2 x

2 x
2 x

1 x
2 x
1 x
1 x

8
10

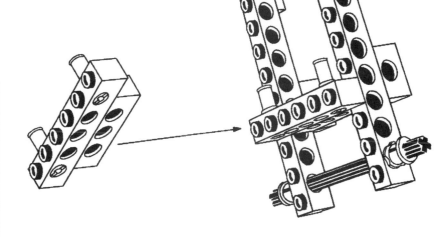

© 1994 LEGO Group

Paul Bunyan's
Bear-Barrow

Paul Bunyan was a friend to all the creatures of the forest. He looked out for their welfare and helped them whenever he could. When forest fires threatened a group of bears, Paul was right there to help them get to safety. He made a bus based on the design of a wheelbarrow. With this, he could easily and quickly get as many as 28 bears to safety at one time.

During one such rescue, eight bears were trying to solve a problem. The bears knew that even Paul's strength was not without its limits. They wished to sit on the bus in such a way that it would be easiest for him to lift.

Help the eight bears to pick the best places to sit!

Build the model and attach the seating pattern. Use it to test out your ideas.

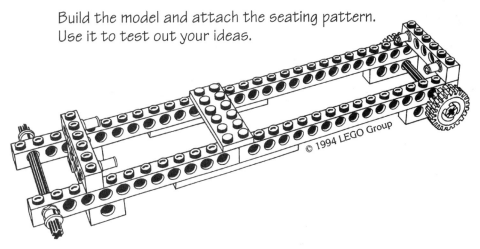

© 1994 LEGO Group

Paul Bunyan's
Bear-Barrow

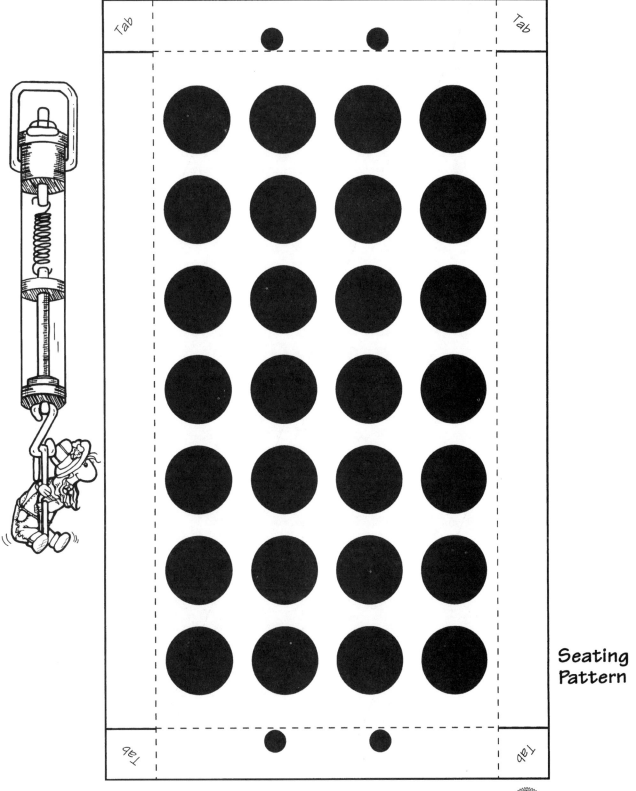

Tab

Tab

Seating
Pattern

Tab

Tab

Paul Bunyan's
Bear-Barrow

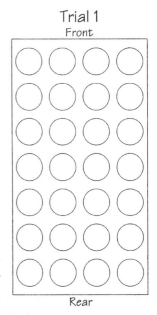

Record the seating arrangements you tried and the predicted force you think is needed to lift the bear-barrow. Use a spring scale to determine the actual effort force needed to lift each.

Trial 1	Trial 2	Trial 3
Front	Front	Front

Predicted Force: _____   Predicted Force: _____   Predicted Force: _____

Actual Force: _____   Actual Force: _____   Actual Force: _____

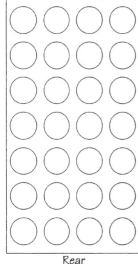

Write a statement that summarizes your findings.

© 1996 AIMS Education Foundation

Paul Bunyan's
Bear-Barrow Challenge

Paul Bunyan was a friend to all the creatures of the forest. He looked out for their welfare and helped them whenever he could. When forest fires threatened a group of bears, Paul was right there to help them get to safety. He built a bus based on the design of a wheelbarrow. With this, he could easily and quickly get as many as 12 bears to safety at one time.

During one such rescue, some of the bears were trying to solve a problem. The bears knew that even Paul's strength was not without its limits. They wished to design a "bear-barrow" in such a way that it would be easiest for him to lift. They decided to have a contest. This notice was posted throughout the forest and sent to several schools. The bears are looking forward to your entry in the contest.

Attention Barrow Making Contest

You are challenged to make a bus of wheelbarrow type design that will carry 12 bears. Each bear must have its own seat.

The winning design is the one that requires the least effort to lift the barrow and 12 bears.

On the back of this paper:
1. Draw your bear-barrow.
2. Record the force needed to raise the handles. (Use a spring scale.)
3. Calculate the mechanical advantage.

A Balance Beam

Topic
Simple Machines: Levers

Key Question
How could you make and calibrate a balance from a simple beam?

Focus
In this activity students will construct a balance and calibrate it to measure the mass of objects.

Guiding Documents
NCTM Standards
- *Extend their understanding of the process of measurement*
- *Select appropriate units and tools to measure to the degree of accuracy required in a particular situation*

Project 2061 Benchmarks
- *In the absence of retarding forces such as friction, an object will keep its direction of motion and its speed. Whenever an object is seen to speed up, slow down, or change direction, it can be assumed that an unbalanced force is acting on it.*
- *A system may stay the same because nothing is happening or because things are happening but exactly counterbalance one another.*

Math
Measuring
Identifying patterns
Using and applying formulae
Using computation

Science
Physical science
   simple machines
     levers

Integrated Processes
Observing
Collecting and recording data
Interpreting data
Applying
Predicting

Materials
LEGO® elements (per group):
- 10 1 x 2 beams
- 10 1 x 4 beams
- 4   1 x 16 beams
- 2   1 x 8 beams
- 1   axle, 6-studs long
- 3   wheels with tires
- 1   weighted brick
- 1   base plate
- 2   2 x 6 plates
- 1   1 x 6 plate
- 2   2 x 4 plates
- 1   pulley wheel
- 1   paper cup
- 1   connector peg

String
Masking tape
Balance
Masses or centicubes

Background Information
The lever balance in this activity is an example of a first-class lever. The fulcrum is located between the effort arm and the resistance arm. The law of the lever states that a lever will be balanced when the product of the force applied to one side of the lever and the distance that force is from the fulcrum is equal to the force of the other side multiplied by its distance from the fulcrum. In the lever balance found in this activity, the effort's force increases as more masses are added to the cup. To balance, the increase in force must be matched by increasing the length of the resistance arm.

Hanging balances of similar design were often used to weigh agricultural products such as bales of cotton.

Management
The following approach is offered for those students ready for more independent exploration.

> *Open-ended:* Provide the students with the drawing of the balance. Ask the *Key Question* and allow students time to develop their solution of making a balance.

1. This activity should take about an hour to complete.
2. Groups of two work well for this activity.
3. The precision to which this balance will measure is limited. It will require patience on the students' part to balance the lever.

BRICK LAYERS   43

4. *Fiddling with Fulcrums* should be done prior to this activity. If students have not done *Fiddling with Fulcrums*, eliminate *Procedure 4*.

Procedure

1. Distribute the construction illustration and have the students make the lever balance.
2. Discuss the *Key Question*.
3. Have students use the string to hang a cup from the end of the beam and place a strip of masking tape down the top length of the beam.
4. Review the findings from *Fiddling with Fulcrums*. In particular, discuss the results of the graph. Have students predict what the marks on the balance will look like. Will they be evenly spaced?
5. Direct the students to slide the beam onto the fulcrum (tires) and balance it with the empty cup hanging over the edge of a table.
6. Instruct them to make a mark on the masking tape to indicate the position of the fulcrum under the beam.
7. Direct the students to add a five-gram mass to the cup and move the beam until it is balanced. Have them place another mark on the tape to indicate this new position.
8. Tell the students to repeat this procedure as many times as possible.
9. Guide them in how to use the lever balance to determine the mass of several objects. Urge them to check its accuracy with a school balance.

Optional:

10. For greater masses, the beam can be turned over. The weighted mass can be attached to the free end of the beam with two 2x4 plates.
11. The beam is balanced on the fulcrum and a new calibration is done starting with 50 grams.

Discussion

1. What did you need to do to counter the added mass placed in the cup? [make the beam longer on the other side of the fulcrum]
2. What pattern do you notice about marks on beam? [The marks get closer together as the mass increases.]
3. What limit is there to a balance of this design? [The marks get too close to be distinguishable.]
4. What could you do to make this balance more precise? [Answers will vary. One suggestion: use a lighter, thinner beam so it with take more of its length to counter the opposing mass.]
5. When was your beam most accurate? [at lower masses]

Extension

Replace the LEGO® beam with a thin wood strip or ruler. Calibrate it. This beam can now be taken home and used with any fulcrum to measure the mass of objects.

A Balance Beam

Parts Inventory

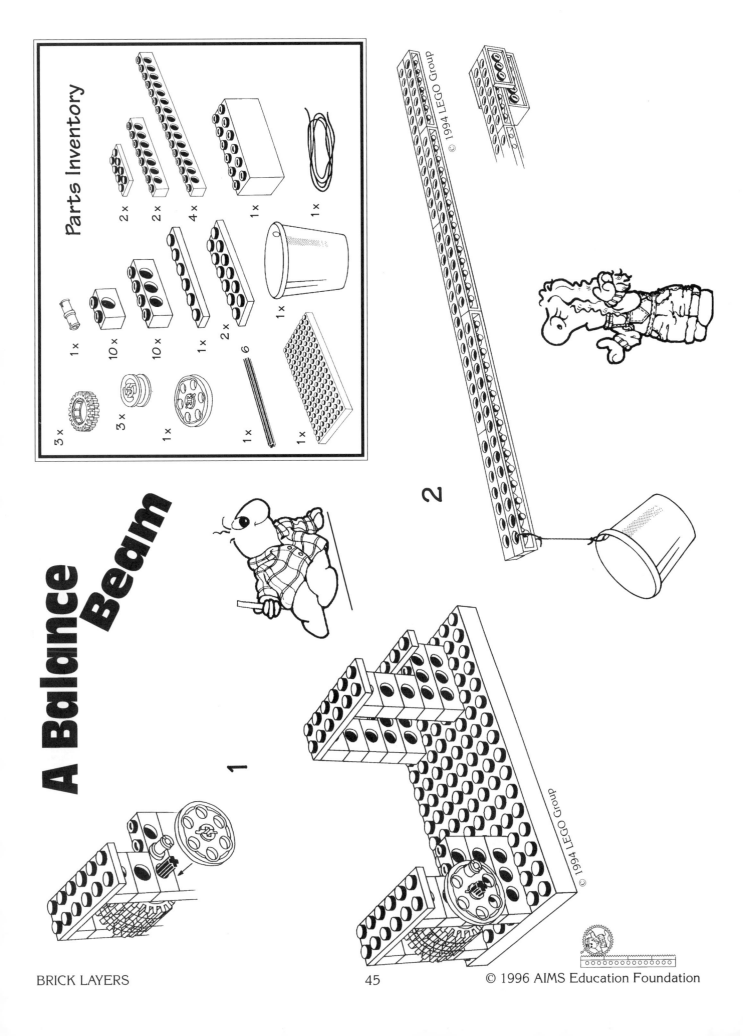

- 2x
- 2x
- 4x
- 1x
- 1x
- 1x
- 10x
- 10x
- 1x
- 2x
- 1x
- 3x
- 3x
- 1x
- 1x
- 6
- 1x
- 1x

1

2

© 1994 LEGO Group

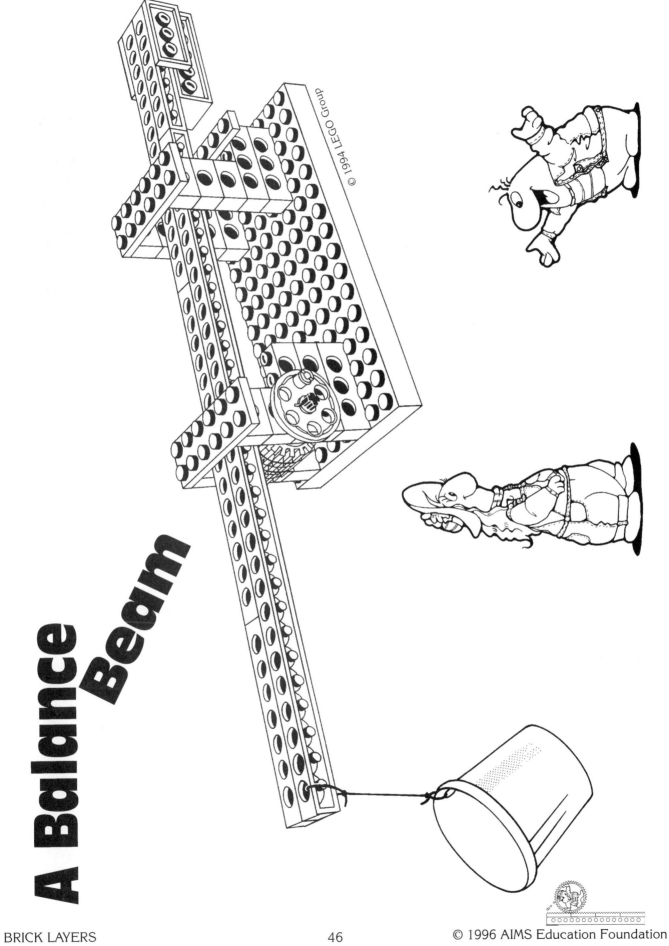

A Balance Beam

46

Wheel and Axle

The wheel and axle is a simple machine. Not only did someone have the great idea to make a wheel round so it would roll easily, but it is also a machine that exchanges force for distance or distance for force—like all other simple machines.

One way to understand the wheel and axle is to think of it as a lever. In this illustration, notice that to balance the boy, the child has to sit twice as far from the seesaw's fulcrum as the boy. This is because the child's weight is one-half that of the boy's. Sitting twice as far from the fulcrum doubles the child's force, making it possible for him to balance a boy twice his weight. However, when they start playing, the child will move up and down twice as far as the boy.

If the seesaw were turned around on the fulcrum, it would make a merry-go-round. The child can easily spin the ride and the boy. Each time they go around, the child will go twice as far as the boy.

To change this lever into a wheel and axle, all that needs to be done is to make the board into a circle. Now it is like a lever extending in all directions from the fulcrum. The child makes the ride go around by using a little force and going a long distance. The boy makes the ride turn by using a large force and going a small distance. Both the boy and the child do the same amount of work to turn the merry-go-round, they just get it done in different ways.

Because of where they sit on the merry-go-round, the boy and child can turn the ride with the same force. The boy is twice as strong as the child, but he pushes at one-half the distance from the turning axis (fulcrum). The child with one-half the boy's strength is twice as far from the axis. They both make the merry-go-round turn with the same force. This turning force is called torque. More force can be made by pushing harder or by being farther from the turning axis.

A car uses the wheel and axle to make it go fast. The strong car engine turns a small axle with a lot of force but one turn is not very far around. The wheel connected to the axle pushes on the road with less force than the engine pushes the axle because the wheel is farther from the turning axis. When the wheel turns once, it goes a lot farther than the axle goes in one turn. The same amount of work is being done at the axle and the wheel; it's just being done in different ways. At the axle there is a big force turning a little distance. At the wheel there is a smaller force turning a greater distance. The wheel and axle let the car trade the big force of the engine for the long distance of a turning wheel.

A winch uses the wheel and axle to make it easier to pick up heavy things. The big crank is turned by a small force around the long distance of one rotation. This turns the little axle one time with a big force. The winch trades a wimpy force for a strong force by going a long distance to get a little distance.

Wheels and axles come in many different forms. Look at the illustrations and try to identify the wheel and axle in each. Try to determine how the force and distance are traded.

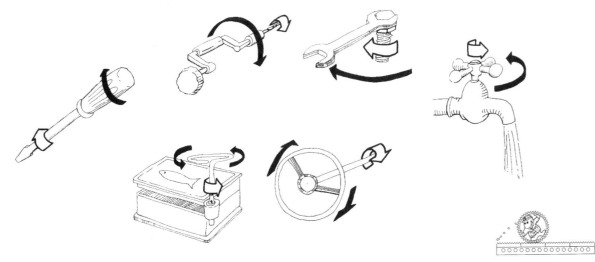

WHEELING Your Way to the TOP

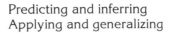

Topic
Simple Machines: Wheel and Axle

Key Question
How does the size of the wheel on a winch affect the rate at which it lifts its load?

Focus
Students will construct a winch using three different wheel sizes. They will crank the winch and measure the amount of string used by each wheel in a rotation to establish the relationship of wheel size to rate of lift.

Guiding Documents
NCTM Standards
- *Explore problems and describe results using graphical, numerical, physical, algebraic, and verbal mathematical models or representations*
- *Describe and represent relationships with tables, graphs, and rules*
- *Analyze functional relationships to explain how a change in one quantity results in a change in another*

Project 2061 Benchmarks
- *Graphs can show a variety of possible relationships between two variables. As one variable increases uniformly, the other may do one of the following: always keep the same proportion to the first, increase or decrease steadily, increase or decrease faster and faster, get closer and closer to some limiting value, reach some intermediate maximum or minimum, alternately increase and decrease indefinitely, increase and decrease in steps, or do something different from any of these.*

Math
Measuring
   length
Graphing
   slope
   extrapolating
Averaging
Using formulae
Using proportional reasoning
Identifying patterns
Using computation

Science
Physical science
   simple machines
      wheel and axle

Integrated Processes
Observing
Collecting and recording data
Interpreting data

Predicting and inferring
Applying and generalizing

Materials
LEGO® elements (per group):

1	building plate
2	1 x 16 beams
2	1 x 12 beams
2	1 x 8 beams
1	1 x 6 beam
6	1 x 4 beams
3	2 x 4 bricks
16	connector pegs
1	steering wheel
2	gears, 40-tooth
2	pulley wheels
3	bushings
1	piston rod
1	axle, 10-studs long

String
Metric tape measure or meter stick

Background Information
Winches are used in many machines. They are used in the hoist on a tow truck and in high-speed elevators. Winches are based on the simple machine called a wheel and axle. As with all simple machines, the wheel and axle makes it possible to exchange force for distance or distance for force. This activity will look at the change of distance by measuring the amount of string used in each rotation of the winch.

The students will make winches by assembling three wheels of various sizes along one axle. As the axle turns, all three wheels turn. When the axle is turned one full turn, all three wheels will make one complete rotation. However, points marked on the circumferences of each of the three wheels will vary greatly in the distances traveled to make one rotation. The distance the points move is directly related to the diameters of the wheels. For example, if one wheel is twice the diameter of a second wheel, a point on the circumference of the first wheel will move twice the distance of a point on the circumference of the second wheel.

In this activity, the winch keeps the rotations of wheels the same, making the measure of distance quite easy. By rotating the axle to which all the wheels are attached, students can see the wheels' different rates of lift.

This experience provides an opportunity for students to see how a graph can be a representation of a physical event. As students make three broken-lines on the graph, they will discover that the steepness of the lines represents the rates at which the wheels lift their loads. The steeper the line, the faster the wheel accom-

plishes its lift. This steepness is referred to as the *slope* of the line. The line rises a certain amount for each rotation of the axle. The height of the line at any point is the sum of the distances of rises for all the preceding rotations. Students will recognize that the lines are relatively straight. This means the rise for each rotation is about the same, or proportional. The average rise for all rotations tells how much one could expect the strings to rise on each rotation. The average rise is a measure of the wheel's rate at which it lifts its load. *The average rise per rotation is the slope of the line.*

To predict the lift for a given number of rotations, students can extrapolate (extend the line on the graph past the points of data) or multiply the average rise per rotation by the given number of rotations. Using the last method, students often find that when they apply their calculated answers to actual results, there is some degree of error. The error often results because the loads were at different levels from the floor at the beginning of the activity. If this is the case, the error will be the distance their loads were from the floor. The beginning height of the load (the distance from the floor) will need to be added to get a more accurate answer. To make a generalized statement of the height of the load, students need to follow this equation:

Final height = (average rise per rotation) x (number of rotations) + (starting height)

Management

1. This activity is best done in groups of two. During the measurement section of the activity, one student is assigned the job of turning and holding the crank. The other student measures and records the data.
2. Students may have difficulty attaching the string to the smallest wheel. Have them tie a loop at the end of the string and put the loop on the axle between the bushings. Have the students push the bushings together to hold the string. The students should wrap the string around the wheel several times before assembling the winch.
3. Determine what skills involved in this activity are appropriate for your students. It may not be appropriate for lower-grade students to calculate the slope of each line and write an equation for the line. They can be asked to make a correlation between the steepness of the line and how quickly the brick was lifted. They can use the graph to predict future events. Another adaptation that will help lower-grade students is to have them adjust the bricks on their lines before starting so the bricks are touching the floor and the line is taut.

Procedure

1. Discuss what a winch is and where it is used. Discuss the *Key Question*.
2. Have students assemble the winch as shown. During the construction of the winch, have them attach a string to each wheel that is long enough to

reach the floor from the table top. Have them attach a LEGO® brick to the loose end of each string.
3. Direct the students to place the winch on the edge of the table so the strings hang to the floor. Have them rotate the crank until all the bricks are lifted off the floor. This is the starting position. (If appropriate, direct students to adjust the bricks so they are all resting on the floor.
4. Have the students measure how far each brick is off the ground and record the distances on the chart. (If the bricks are resting on the floor, tell students to record 0 cm on the chart.)
5. Direct the students to turn the crank one rotation.
6. Have the students measure and record the heights of the bricks from the floor to their new positions.
7. Instruct the students to continue in this manner, turning the crank one rotation and calculating how far each brick has risen from the last position until the brick on the large wheel can rise no more. Have them record these calculations on the chart under *Rise from last turn.*
8. Have students make a graph of their data and discuss their results.

Discussion

The science concept that larger wheels lift a load at a faster rate is easily grasped by students and is quite evident; however, a mathematical understanding is more complicated. The *Background Information* gives a more thorough discussion of the development of the concept of graphing and writing an equation. The questions below provide an outline of what questions might be asked to help students develop the mathematical concepts. The questions are in order of conceptual development so the teacher can decide what is appropriate for the students.

1. How do the lines on the graph show what happened when you turned the crank? [The faster the lift, the steeper the line.]
2. How could you use the graph to predict how high each brick would be after nine rotations? [Extend the lines of the graph and read the distance for nine turns.]
3. How could you use the *Average Rise Per Turn* to calculate how high a brick would be after nine turns? [Add the average rise nine times or multiply the average rises by nine.]
4. How could you use the *Average Rise Per Turn* to write a generalization of how high the load would rise if you were given a certain number of rotations? [height = (average rise) x (rotations) + (starting height) Few students will add the starting height. If this is the case, have them check to see if their equations work. By asking them to refer to the graph and recall what they did, most students will recognize the need to make the addition.]

Extension

Have students explore the relation of the wheel's diameter and circumference to the rate of lifting.

WHEELING
Your Way to the
TOP

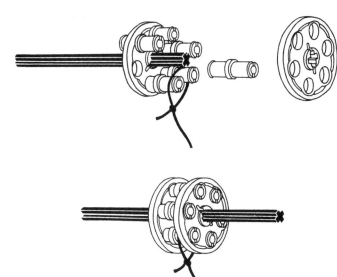

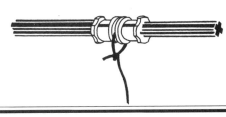

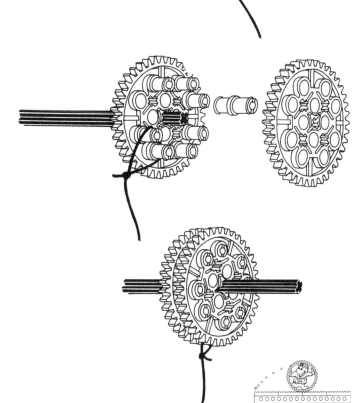

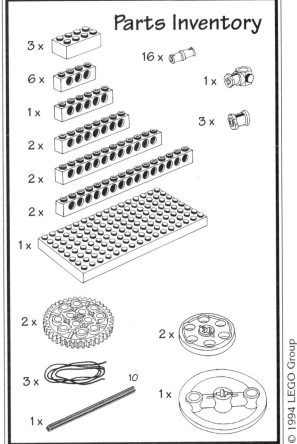

Parts Inventory

3 x
6 x
1 x
2 x
2 x
2 x
1 x

16 x
1 x
3 x

2 x
2 x
3 x
10
1 x

© 1994 LEGO Group

51

WHEELING Your Way to the **TOP**

© 1994 LEGO Group

WHEELING Your Way to the TOP

How does the size of a wheel affect the rate at which it lifts its load?

Build the winch as shown on the assembly sheet. Attach a brick (the load) to the end of each string. Measure and record the distance from the floor to the bottom of the load. Turn the crank one full rotation. Measure and record the distance from the floor to the bottom of the loads again. Continue this process until one of the strings cannot be raised any more. Complete the chart by finding how much each string rose with each turn.

Turns of the crank	Small Wheel		Medium Wheel		Large Wheel	
	Distance off the floor (cm)	Rise from last turn (cm)	Distance off the floor (cm)	Rise from last turn (cm)	Distance off the floor (cm)	Rise from last turn (cm)
Start 0		▓		▓		▓
1						
2						
3						
4						
5						
	Average Rise Per Turn		Average Rise Per Turn		Average Rise Per Turn	

BRICK LAYERS

53

WHEELING
Your Way to the TOP

Make a broken-line graph for each of the three sizes of wheels. Use a different color for each line. Include a key to explain the colors.

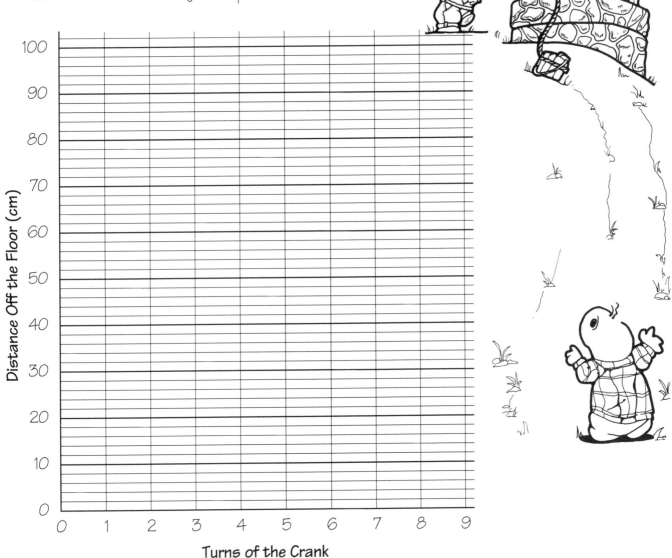

Distance Off the Floor (cm)

100
90
80
70
60
50
40
30
20
10
0

0 1 2 3 4 5 6 7 8 9

Turns of the Crank

1. How do the lines on the graph show what happened when you turned the crank?
2. How could you use the graph to predict how high each string would be after nine turns?
3. What does the *Average Rise Per Turn* tell you about the graph?
4. Write an equation that tells how far from the floor a load will be if you turn the crank a given number of times.

BRICK LAYERS   54   © 1996 AIMS Education Foundation

A SHIFT IN LIFT

Topic
Simple Machines: Wheel and Axle

Key Question
What happens to the force you apply to a winch if you change the size of wheel you turn?

Focus
Students will construct a winch with a winding wheel on each end. While keeping one wheel the same size, they will change the wheel size on the other end. Students will hang masses on each wheel to keep them in balance, and determine how the wheel size affects the force generated by the wheel. Using the data they gather, students will make broken-line graphs to visualize the information and help in constructing an equation.

Guiding Documents
NCTM Standards
- *Explore problems and describe results using graphical, numerical, physical, algebraic, and verbal models or representations*
- *Represent situations and number patterns with tables, graphs, verbal rules and equations and explore the interrelationships of these representations*

Project 2061 Benchmarks
- *In the absence of retarding forces such as friction, an object will keep its direction of motion and its speed. Whenever an object is seen to speed up, slow down, or change direction, it can be assumed that an unbalanced force is acting on it.*
- *Thinking about things as systems means looking for how every part relates to others. The output from one part of a system (which can include material, energy, or information) can become the input to other parts. Such feedback can serve to control what goes on in the system as a whole.*

Math
Measuring
Graphing
Averaging
Identifying patterns
Using rational numbers
Using computation

Science
Physical science
    simple machines
        wheel and axle
        torque

Integrated Processes
Observing
Collecting and recording data
Interpreting data
Controlling variables
Predicting and inferring
Applying and generalizing
Comparing and contrasting

Materials
LEGO® elements (per group):
1	building plate
2	1 x 4 beams
1	axle, 12-studs long
18	connector pegs
2	40-tooth gears
2	24-tooth gears
2	pulley wheels
2	beveled gears
1	bushing

String
Paper cups
Centicubes

Background Information
In terms of force, a wheel acts as a lever. The greater the wheel size, the greater the distance from the axle (the fulcrum) the force is applied. The greater the distance from the fulcrum the force is applied, the greater the *torque,* or turning force, that is generated.

For two wheels of different sizes to generate equal amounts of torque, more force must be applied to the smaller wheel. In an ideal situation, the amount of force required on the smaller wheel is an inverse relationship to the size of the wheels. A wheel that is half the size of the other wheel will require twice the amount of force to balance the torque generated by the larger wheel.

The winch built in this activity is not an ideal situation; friction interferes and the actual outcomes will not precisely follow the theoretical ideal.

The use of a graph helps students to visualize the relationship of the wheel sizes to the forces they generate. It also provides a developmental opportunity for students to learn to write equations about graphs. For more information about this development, see *Background Information* of *Wheeling Your Way to the Top.*

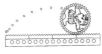

Management

1. This activity is best done in groups of two. During the measurement section of the activity, one student should hold the winch to the table and record the data while the other student balances the cups of centicubes.

2. Students may have difficulty attaching the string to the smallest wheel. Have them tie a loop at one end of the string and put the loop on the axle between the bushing and one of the gears. Have the students push the bushing and gear together to hold the string.

3. Because of the amount of fiction and the small masses used in this activity, students will have to take some time in determining if the masses are balanced. Before beginning the activity, carefully read the balancing procedure in step 9 to be sure the proper procedure can be communicated to the students so they can obtain reasonable results.

4. After the students assemble the winch, they need to be given adequate time to discover how this winch works. Pose the problem, "When you pull on one line of a winch, the other line raises. How can you get this winch to work?" Leave the problem open-ended, providing for free exploration. When a number of students have found the solution, they should be asked to discuss their findings before beginning the quantitative part of the activity. If free exploration is not appropriate for the class, begin by explaining how to work the winch. In either case, students will need to be given some time to explore the winch.

5. The strings on the wheels will wind in opposite directions so that one is clockwise and the other is counterclockwise. Wind one string leaving the other string unwound. Pulling down on the wound string will cause the other string to rise.

6. To keep the string in line with the wheels and clear of objects while it winds up, students will need to hold the base of the model on the table so the wheels of the winch are cantilevered past the table's edge.

7. Because of friction, the data gathered in this activity are not extremely consistent. Students will need to realize the equation they generate is developed from a line of best fit.

Procedure

1. Discuss the *Key Question*.
2. Direct students to assemble the winch as shown in the illustration.
3. Have them cut two strings long enough to reach from the floor to the top of the table on which they are working. Direct them to tie a small loop in one end of the string. Inform students that as each winding wheel is assembled and attached to the winch, they need to use this loop to attach it to the wheel.
4. Guide the students to attach a medium-sized wheel

to one side of the winch and the large wheel to the other side.

5. Have them punch a hole near the rim of each of the two paper cups. Direct students to tie the cups to the strings.

6. Explain that they need to place the winch on the edge of the table so the cups reach the floor.

7. Allow the students time to discover how to get the winch to work. Have them discuss their strategies before beginning the quantitative part of the activity that follows. (Refer to *Management 4, 5, and 6*.)

8. Tell the students to adjust the wheels so the cup on the medium-sized wheel (the control) is raised to the top and the opposing cup is down on the floor.

9. Have the students hold the cup that is attached to the large wheel on the floor. Direct them to place the two centicubes (two grams) in the medium-size wheel's cup (control). Inform the students to release the cup on the floor. If the control cup drops when the other cup is released, have them raise it again and add one centicube to the cup on the floor. Direct the students to again release the cup on the floor. If the control drops again, have them add more mass, a centicube at a time, until the control cup does not drop. When the control cup does not drop, urge students to alter its position to test whether it will move in the new position. If it does not move in any position, the two cups are considered balanced. If the control cup moves in other positions, the amount of centicubes in the other cup needs to be adjusted until the control cup does not move.

10. When students have determined that the cups are balanced, have them determine and record the number of centicubes balancing the control side.

11. Direct them to continue this procedure using the other quantities listed under *Centicubes on Control Side*.

12. Have students calculate the increase of centicubes used to balance the control side each time mass is added and record this increase on the chart.

13. Have them calculate the *Average Increase* of centicubes for each two centicubes added to the control side.

14. Have them replace the large wheel with the medium-sized wheel and repeat the procedures. When the data have been collected with the medium wheel, have them replace it with the small wheel and collect data for it.

15. Direct students to complete the graph and discuss their observations.

Discussion

1. What happened to the number of centicubes used to balance the control cup as the size of the wheel increased? [It takes fewer centicubes the larger the wheel gets.]

2. How does the graph show that fewer centicubes are needed to balance the control cup when the larger wheel is used? [The larger the wheel, the less steep the line, or its slope is less.]

3. Ask questions such as: How could you use the graph to predict how many centicubes would be used to balance 15 centicubes in the control cup with the large wheel? [Extend the large wheel line and go horizontally from its intersection with the 15 centicubes in the control cup to determine the centicubes needed to balance. (Solutions will vary.)]

4. Ask questions such as: How could you use the *Average Increase* to calculate how much it would take to balance 15 centicubes in the control cup with the large wheels? [The *Average Increase* is for two centicubes, so divide the *Average Increase* by two to determine the average increase per centicube. Multiply the average increase per centicube by 15.]

5. What does the *Average Increase* tell you about its corresponding line on the graph? [how far it goes up for each two squares it goes horizontally; The greater the average increase, the steeper the line (slope).]

6. Use the *Average Increase* to write an equation for each line. Choose a number to use as the centicubes used in the control and see if your equation gives you roughly the same answer you get on the graph. (You may need to draw attention to the fact that the *Average Increase* is for each two centicubes added to the control side.) [Possible equations are:
   Balancing Centicubes = (Average Increase/ 2) x Number of Centicubes in Control
   Balancing Centicubes = Average Increase x (Number of Centicubes in Control/2)]

7. How would you design a winch if you had to lift a load that is much heavier than you can lift with no other help? What size winding wheels would you use? [Attach the load to a small wheel, and turn a large wheel.]

8. What similarities and differences do you see between this graph and the one you made in *Wheeling Your Way to the Top*? [Answers will vary, but students should mention that both graphs have all lines that slope upward to the right (positive slope) and that the arrangement of wheel size is opposite (speed and force have inverse relationships).]

Extensions

1. Repeat the activity using the small wheel and large wheels as controls.

2. Use a balanced system to calculate the amount of work done on each side of the winch. Centicubes multiplied by distance cup moves on the control side ideally should equal centicubes multiplied by distance cup moves on the other side. Because of friction this will not be true, providing an opportunity to discuss efficiency.

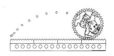

A SHIFT IN LIFT

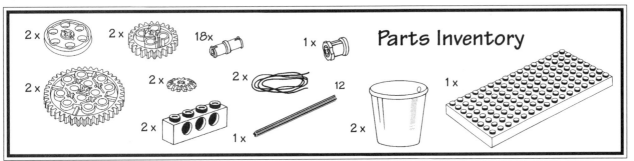

Parts Inventory

2 x
2 x
18x
1 x

2 x
2 x
2 x
12
1 x

2 x
1 x
2 x

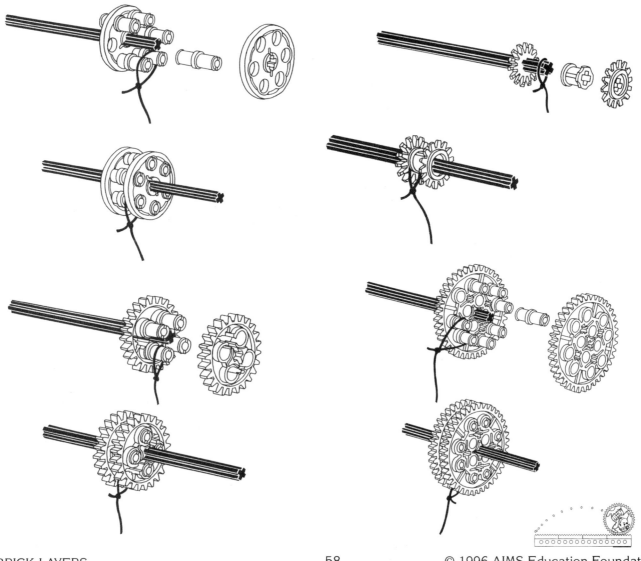

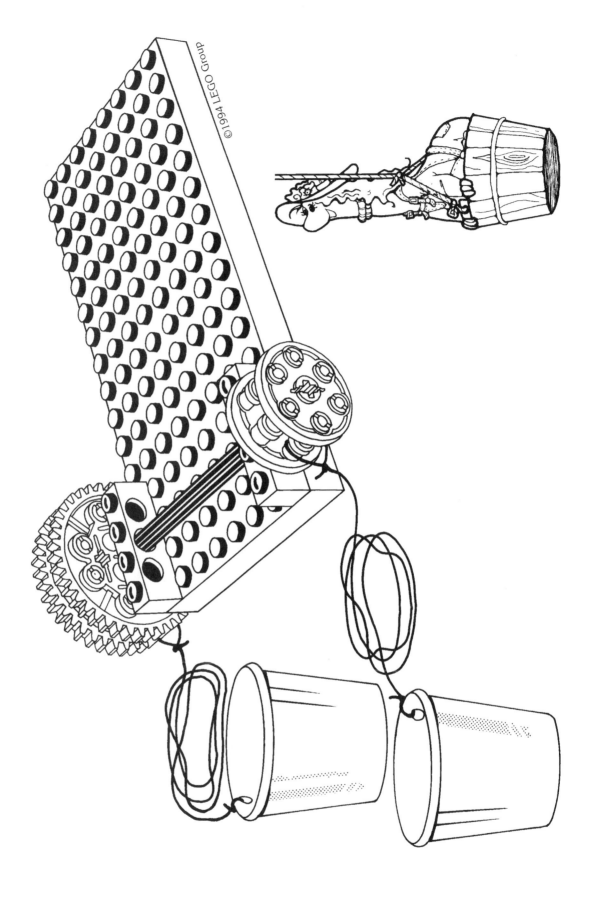

A SHIFT IN LIFT

©1994 LEGO Group

A SHIFT IN LIFT

If you change the size of the wheel you turn, what happens to the amount of force you need to apply?

Build the winch as shown. Use the medium-sized wheel as the control wheel and place the correct number of centicubes in its cup. Add centicubes to the other wheel's cup until the cups do not move, indicating they are balanced. Record the number of centicubes put in the other wheel's cup. Continue this procedure for the different control masses and different-sized wheels.

Centicubes on Control Side	Large Wheel		Medium Wheel		Small Wheel	
	Centicubes Balancing Control	Increase in Balancing Centicubes	Centicubes Balancing Control	Increase in Balancing Centicubes	Centicubes Balancing Control	Increase in Balancing Centicubes
	Average Increase		Average Increase		Average Increase	

A SHIFT IN LIFT

Make a broken-line graph for each of the three wheels. Make each line a different color and label each line.

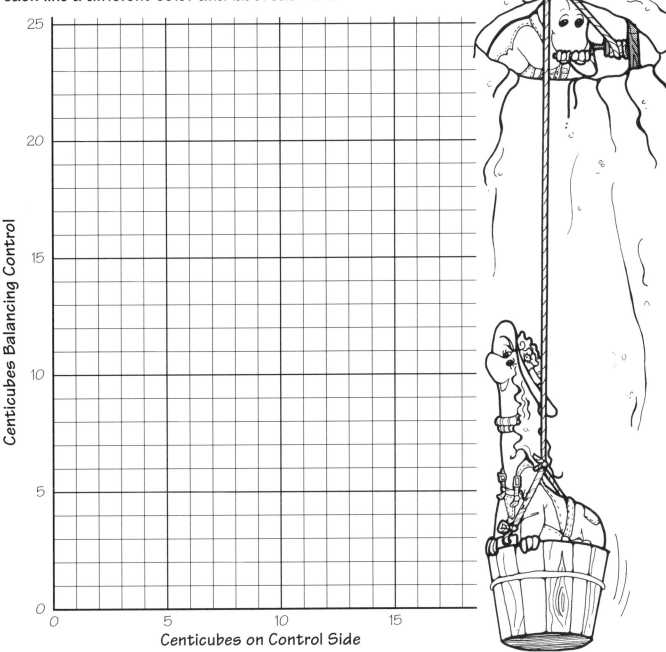

1. How is the average increase for each wheel represented on the graph?

2. Use the *Average Increase* to help you write an equation for each line. If you know the number of centicubes on the control side, the equation should be able to tell you how many centicubes to put into the other side to balance the control.

SLOT CARS

Topic
Simple Machines: Wheel and Axle

Key Question
How does the size of a slot car's drive wheel affect its performance?

Focus
Students construct a car which is run by a falling mass. By changing the size of the drive wheel, the distance and the rate the car travels change.

Guiding Documents
NCTM Standards
- *Systematically collect, organize, and describe data*
- *Represent numerical relationships in one- and two-dimensional graphs*

Project 2061 Benchmarks
- *Energy can change from one form to another, although in the process some energy is always converted to heat. Some systems transform energy with less loss of heat than others.*

Math
Measuring
Graphing
Averaging
Using computation

Science
Physical science
   simple machines
      wheel and axle
      conservation of energy

Integrated Processes
Observing
Collecting and recording data
Interpreting data
Generalizing
Comparing and contrasting

Materials
LEGO® elements (per group):
- 2   1 x 16 beams
- 2   2 x 6 plates
- 2   axles, 8-studs long
- 4   wheels with tires
- 1   bushing
- 2   beveled gears
- 14 connector pegs
- 2   pulley wheels
- 2   gears, 40-tooth
- 1   weighted brick

String
Rubber band
2 wooden meter sticks (per group)
Masking tape

Background Information
A mass suspended in air contains the energy needed to do work, *potential energy*. As the mass drops, the potential energy is changed into motion, *kinetic energy*. In this activity, the weighted LEGO® brick provides the mass to run the slot car. As the mass drops, its kinetic energy will be transferred to a wheel by a string that connects the brick and the wheel. For each trial, the car will receive the same amount of energy because the same mass will fall the same distance.

With the input energy the same for each trial, the same amount of *work* is done. The two components of work, *distance* and *force*, should be the same if the car is not changed. To obtain different distances and rates (the result of force), the size of the drive wheel on the car must be changed.

In this activity, the car with the largest drive wheel will go the shortest distance, but will move at the fastest rate. The car with the smallest drive wheel will go the longest distance, but will do so at the slowest rate. The car with the middle-sized wheel will fall between these two extremes.

Each car receives an equal amount of energy of motion, but they will differ in how they transform that energy. The radius of the large wheel allows it to generate a large *torque*, or turning force, resulting in very rapid rotations of the wheels. However, the large wheel size has a large circumference, so the string can go around very few times, resulting in very few rotations. The large drive wheel provides a very fast rate for a very short distance. The small wheel's radius produces little torque, and a resulting slow rate. Its small circumference allows the string to make many rotations producing a longer trip.

The wheel and axle, like all simple machines, transforms work. Large input (drive) wheels transform

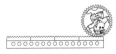

BRICK LAYERS

© 1996 AIMS Education Foundation

the work into a larger force but over a shorter distance. Small input wheels transform the work into smaller forces but over longer distances.

The bar graph is designed to help reinforce students' understanding of the meaning of an average or mean. A bar graph is made for each trial. A line drawn across the three bars represents the average of the three trials. This method allows students to recognize that finding the mean is just a method of leveling the data to find a middle value. Those trials above the mean could give their abundance to those trials that did not reach it.

Management

1. This activity is best done with groups of two students. One student works the car while the other measures and records data.
2. Because students may have difficulty attaching the string to the smallest wheel, have them put the axle through the loop of string with a bushing and beveled gear on either side of the string. Have them squeeze the bushing and gear together to hold the string.
3. The rubber band is tied around the brick as a shock absorber to slow the inertia of the falling brick. Have students make sure the rubber band is secure before they drop the brick.
4. To join the wheels with connector pegs, the center slots for axles on the wheels must be aligned.
5. As the brick reaches the bottom of the drop, it suddenly stops the wheels and the car skids to a stop. To get consistent measurements, have the students measure where the car was at the end of the drop, not the end of the skid.

Procedure

1. Discuss the *Key Question*.
2. Direct students to tie a loop at one end of a 50 cm string.
3. Have them assemble the slot car as shown in the illustration, attaching the loop of string to the drive wheel.
4. Have the students tie a rubber band around the weighted brick and then tie the rubber band to the free end of the string so the brick hangs 30 cm below the car.
5. Guide students to build a track by placing the two meter sticks side by side, so they straddle the space between two tables. Five centimeters of each end of the sticks should rest on each table. Have the students spread the meter sticks apart so the wheels on each side of the car rest in the center of either meter stick. The slot between the meter sticks accommodates the string from which the weighted brick hangs.

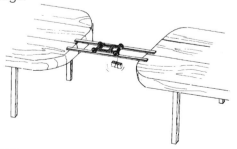

6. Have the students place the car on the meter sticks and wind the string around the drive wheel. When students release the brick, it should drop and provide the energy for the car to cross the track. Allow students time to explore how this car works. Encourage them to adjust the spread of the meter sticks so the car moves freely. Have them use masking tape to hold the meter sticks in the correct position. When students can get their cars to move consistently, have them proceed to the next steps of quantification.
7. Direct the students to wind the string and place the car at one end of the meter stick track. Point out that the brick should be at its maximum height at each release. Have the students release the car and measure the distance it travels. Tell them to perform three trials. Have the students record their data and make a bar graph.
8. Direct them to repeat the above procedure for each size wheel.
9. Ask the students to put the wheels in order by rate.
10. Have them discuss their observations and conclusions.

Discussion

1. What drive wheel should you use to get the longest trip? [Use the smallest wheel.]
2. What drive wheel should you use to get the fastest trip? [Use the largest wheel.]
3. What happens to the distance a car goes as you use larger wheels? [The car goes shorter distances.]
4. What happens to the rate a car goes as you use larger wheels? [The car goes faster.]
5. What is the relationship between the distance and the rate at which the car moves? [The faster a car goes, the shorter the distance it travels. The slower a car goes, the greater the distance it travels.]
6. Explain how the size of a slot car's drive wheel affects its performance. [Answers will vary, but a possible explanation is included in *Background Information*.]

Extensions

1. This activity had students change the size of the drive wheel while keeping the output (traction) wheel the same. Have students change the size of the output (traction) wheels while keeping the drive wheel the same size.
2. With the small wheel on the car, have students measure the distance the car moves and time the duration of a trip. Have them calculate the car's speed.

Parts Inventory

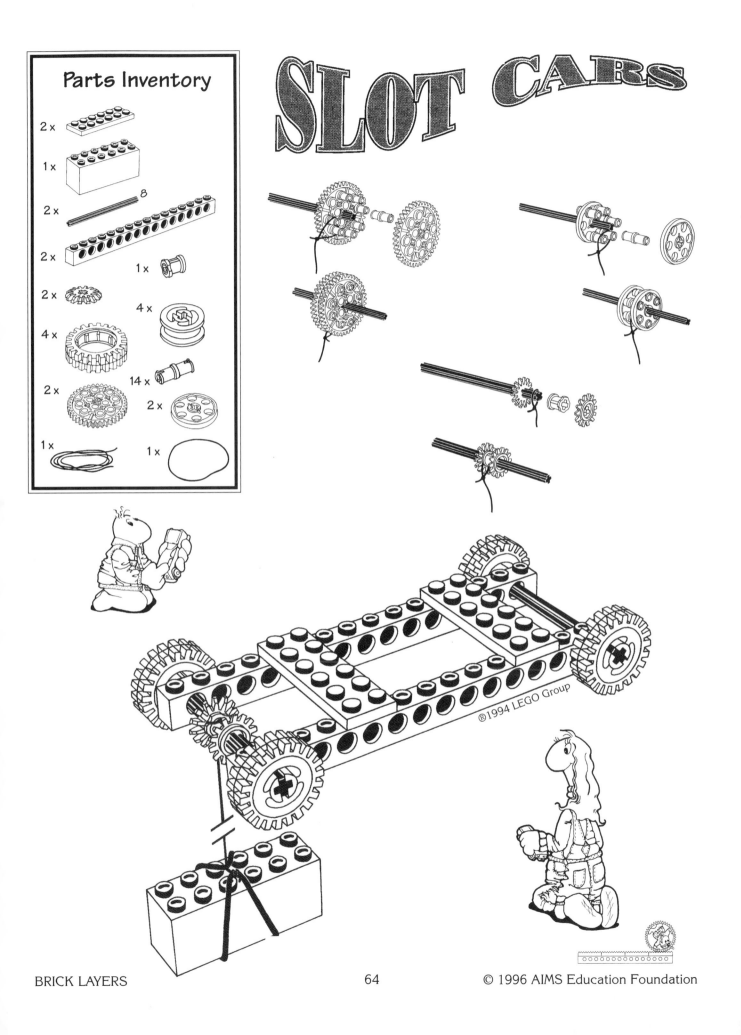

2 x
1 x
2 x — 8
2 x
1 x
2 x
4 x
4 x
14 x
2 x
2 x
1 x
1 x

SLOT CARS

®1994 LEGO Group

SLOT CARS

How does the size of a slot car's drive wheel affect its performance?

1. For each trial, record the distance the car travels and calculate the average.
2. Make a bar graph for each trial. Draw across the bars to indicate the average distance for each wheel size.
3. Order the drive wheel sizes by their effect on the rate of the car.

Distance Car Travels	Small Drive Wheel	Medium Drive Wheel	Large Drive Wheel
Trial 1			
Trial 2			
Trial 3			
Average			

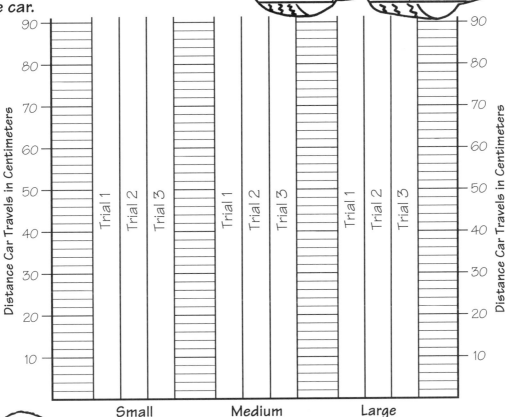

Rate Car Travels

Rate	Fastest	Middle	Slowest
Wheel Size			

LEGO® Launcher

Topic
Simple Machines: Wheels

Key Question
Which design makes the launcher fire the projectile the farthest?

Focus
Students will build launchers with three different gear sizes to determine which gives the longest shot. Students will learn the effect of wheel size to turn rate or distance traveled.

Guiding Documents
NCTM Standards
- *Estimate, make, and use measurements to describe and compare phenomena*
- *Systematically collect, organize, and describe data*
- *Make inferences and convincing arguments that are based on data analysis*

Project 2061 Benchmarks
- *Energy cannot be created or destroyed, but only changed from one form into another.*
- *Energy appears in different forms. Heat energy is in the disorderly motion of molecules and in radiation; chemical energy is in the arrangement of atoms; mechanical energy is in moving bodies or in elastically distorted shapes; and electrical energy is in the attraction or repulsion between charges.*

Math
Measuring
Graphing
Using computation

Science
Physical science
    simple machines
        wheel and axle

Integrated Processes
Observing
Comparing and contrasting
Collecting and recording data
Interpreting data
Generalizing

Materials
LEGO® elements (per group):
  2 40-tooth gears

2	24-tooth gears
2	8-tooth gears
2	gear racks
1	steering wheel
2	1 x 6 plates
2	1 x 8 plates
4	1 x 4 plates
3	1 x 3 plates
2	1 x 16 beams
1	1 x 8 beams
2	1 x 6 beams
4	1 x 4 beams
2	1 x 2 beams
4	2 x 4 bricks
1	axle, 4-studs long
1	axle, 6-studs long
1	bushing

Meter stick

Background Information
A simple machine does not generate any force, it only transforms it. A simple machine changes the amount of a force and the distance that force moves. If a machine increases the force, it reduces the distance that force travels. If a machine decreases the force, it increases the distance that force travels. A simple machine has an inverse effect on the amount of a force and the distance that force travels.

The launchers constructed in this activity use the wheel and axle as a simple machine. The steering wheel is the input where the force is applied for the launch. It is directly connected to the launching gear where the force is output to the projectile. Each time the steering wheel is turned once, the launching gear turns once. With the eight-tooth gear, one rotation of the steering wheel makes the projectile move an eight-tooth length, a 24-tooth length with the 24-tooth gear, and a 40-tooth length with the 40-tooth gear. If the steering wheel is always turned at the same rate, the 40-tooth gear will make the projectile go the longest distance in one turn.

This activity focuses on the change in distance in the simple machine. The change in distance is obvious in the change of speed of the projectile. Not as obvious is the change in force. The LEGO® elements are so light that the change in the force required to move the larger gear is not easily discernible; however, if care is taken, the change in force can also be observed.

This model provides an opportunity to observe a change from the spinning (radial) motion of the gear to

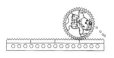

the straight line (linear) motion of the projectile. This type of linkage is very important in technology where a spinning motor must move something back and forth in a straight line, or where a solenoid is used to get something to turn.

Management
1. Two students per group is ideal for this activity.
2. Allow a minimum of an hour for this activity.
3. Students need to be consistent in launching. Encourage them to have the same person in each group spin the wheel so that they can try to spin the wheel at the same rate. They should also start the projectile in the same position. To ensure this, have them move the projectile so it is at the very front of the launcher housing.
4. Measurements must be taken consistently. Have students place the launcher on the edge of a table. Use the meter stick as a plumb to find the location that is directly under the launcher. Students should measure the distance from this location to where the projectile first hits the floor.
5. You may choose to have three sets of partners each build a different launcher so direct comparisons can be made quickly and easily.

Procedure
1. Distribute materials to the students.
2. Have the students construct the launchers following the illustrations.
3. Discuss the *Key Question* and have students make their predictions.
4. Direct the students to launch the projectile six times for each launcher. Have them record these distances on the chart.
5. Have students calculate and record the average distance the projectile traveled for each launcher.
6. Direct them to make a graph from the data. A bar is made for each launch, and the average is drawn as a line across all six bars.
7. Discuss the findings and have students record their generalizations.

Discussion
1. What was the longest launch?... shortest launch?
2. What patterns do you see in the graph? [The bars get shorter as the launch gears get smaller.]
3. What launcher has the longest average launch?...the shortest average launch?
4. What generalization can you make about the average launch and the launch gear? [The larger the launch gear, the longer the average launch.]
5. Why does the largest gear give the longest launch? [It makes the projectile go farther each turn. Going farther each rotation means it goes faster. Going faster means the projectile gains more momentum and goes farther.]
6. How might you modify the launcher to make the projectile go farther?

Extensions
1. Have students repeat the investigation removing the steering wheel and replacing it with a pulley wheel or using just the axle as the input. Have students discuss how the size of the input wheel affects the launch distance.
2. Have students choose the best combination of input wheel and launching gear and construct a launcher with the longest shot. Have students compete in a contest with their launchers.

LEGO® Launcher

Parts Inventory

3 x

2 x

2 x

4 x

2 x

2 x

2 x

1 x

2 x

1 x

1 x   4

1 x   6

1 x

2 x

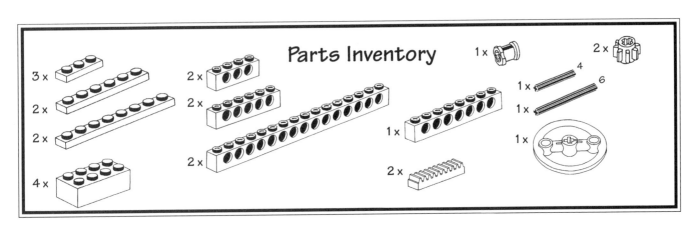

1

© 1994 LEGO Group

© 1994 LEGO Group

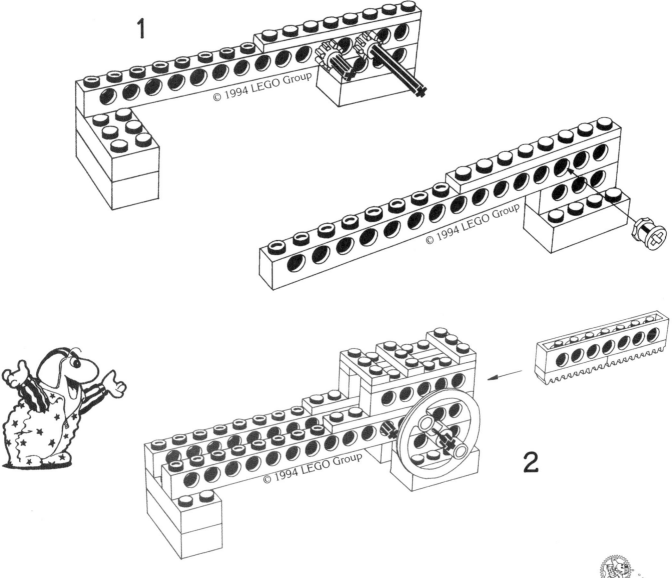

© 1994 LEGO Group

2

LEGO® Launcher

1

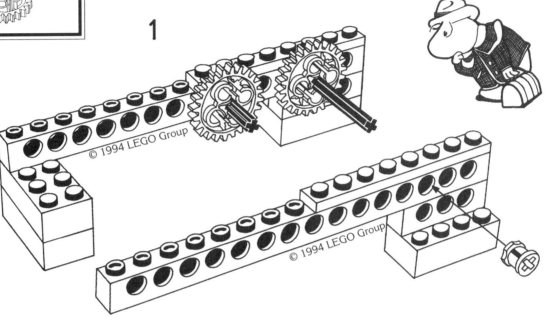

© 1994 LEGO Group

© 1994 LEGO Group

2

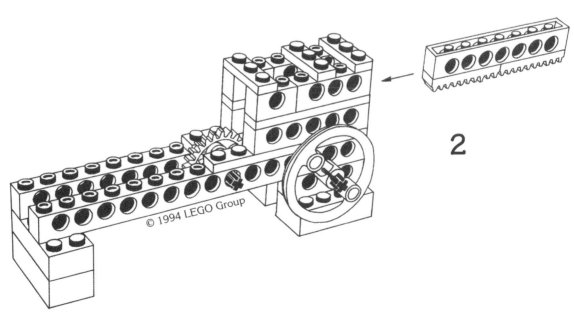

© 1994 LEGO Group

© 1996 AIMS Education Foundation

LEGO® Launcher

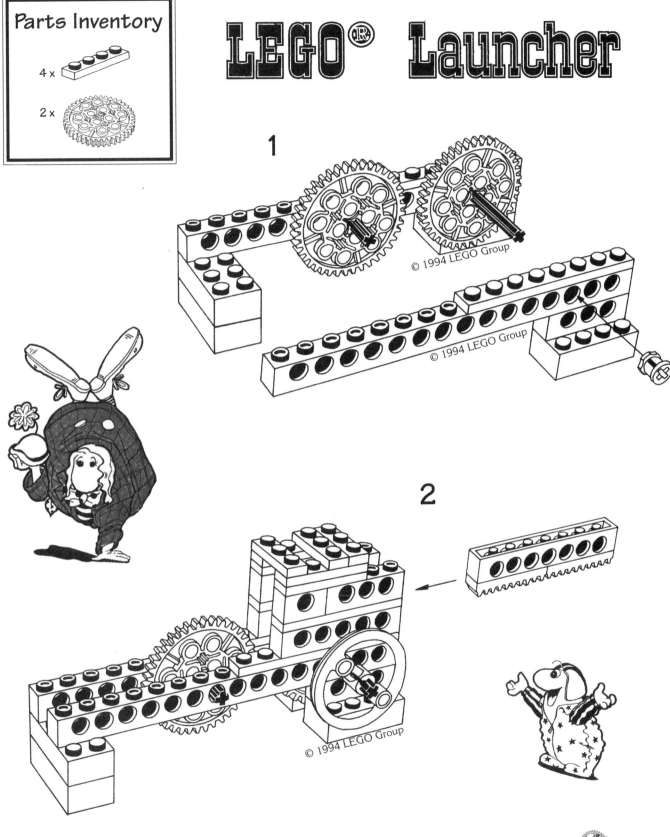

Parts Inventory

4 x

2 x

1

© 1994 LEGO Group

© 1994 LEGO Group

2

© 1994 LEGO Group

LEGO® Launcher

Which design launches the projectile the farthest

Launch the projectile six times with each size launcher. Measure the distance of each launch and record it on the chart below. Then calculate the average distance for each launcher.

	8-Tooth Launcher	24-Tooth Launcher	40-Tooth Launcher
Launch #1			
Launch #2			
Launch #3			
Launch #4			
Launch #5			
Launch #6			
Average			

Make a bar graph of the data with a bar for each launch. Draw a line for the average across all six bars for that size launcher.

Using the information from the chart and the graph, write a generalization about the gear size of the launcher, and the distance the projectile goes.

BRICK LAYERS

71

© 1996 AIMS Education Foundation

LEGO® Launcher

8-Tooth Launcher

24-Tooth Launcher

40-Tooth Launcher

Distance of Launch (cm)

Launch #1 Launch #2 Launch #3 Launch #4 Launch #5 Launch #6

Gears

A gear is a modification of the wheel and axle. It just has teeth around it.

Like all simple machines, gears may change the direction in which a force is applied; or increase/reduce a force or the distance over which a force is applied.

Gears work in teams. Two gears working together is a combination of two simple machines. When two or more machines work together, as in the case of a pair of gears, it is called a *compound machine*.

Two or more gears working together is called a *gear train*. The gear on the train to which the force is first applied is called the *driver*. The final gear on the train to which the force is transferred is called the *driven gear*. Any gears between the driver and driven gears are called *idlers*.

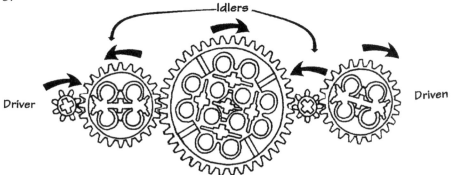

When the teeth of two gears are meshed so that one gear drives the other, the gears turn in opposite directions. If a gear train has five gears, the first, third, and fifth gears will turn the same direction while the second and fourth gears will turn the opposite direction.

Counting the teeth on a gear makes it easy to predict what will happen when you turn the gear. A gear with 40 teeth has five times as many teeth as an eight-tooth gear. If these two gears are meshed and the larger gear is turned once, 40 teeth will drive the smaller gear. The eight-tooth gear will need to turn five times ($8 \times 5 = 40$) to mesh with the one turn of the 40-tooth gear. The small gear has to turn at a faster rate to keep up with the larger gear. The change in the turning rate can be determined by making a ratio of the drive gear to the driven gear. With these gears, the ratio of teeth is 40:8 (40/8) or 5:1 (5). This ratio also tells us that the turning rate of the driven gear is five times faster than that of the drive gear. The drive gear is five times as big, but the driven gear turns five times as fast.

If you were to hold the axle of the 40-tooth drive gear and turn it one time, your fingers would not have to move very far. The axle of the eight-tooth driven gear would go five times the distance as the other axle. This machine is exchanging a small distance for a long distance. Remember: You can't get something for nothing, so there must be an exchange going on. We know part of the exchange because the drive gear's axle is going a small distance and the driven gear's axle is going five times as far:

<u>Driver's Work</u>   <u>Driven's Work</u>

? FORCE X 1 DISTANCE = ? FORCE X 5 DISTANCES

To increase the distance five times, the force has to decrease to make a fair trade. When distance increases by a factor of five, the force becomes one-fifth as strong.

If a rope were connected to each axle so there could be a tug-of-war, which rope would you choose?

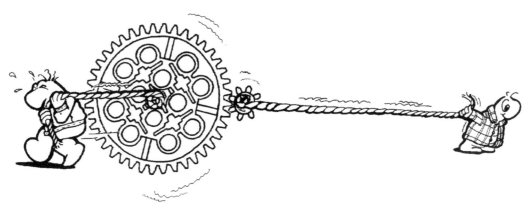

5 FORCES X 1 DISTANCE = 1 FORCE X 5 DISTANCES

The child in the illustration chose the eight-tooth side because it makes his little force five times as great; he just has to be willing to pull five times as far.

Magic String

Topic
Simple Machines: Gears

Key Question
How do the gears make this "magic" trick work?

Focus
Students examine and observe a gear train to determine how it works.

Guiding Documents
NCTM Standards
- *Apply mathematical thinking and modeling to solve problems that arise in other disciplines, such as art, music, psychology, science and business*
- *Describe and represent relationships with tables, graphs, and rules*
- *Estimate, make, and use measurements to describe and compare phenomena*

Project 2061 Benchmarks
- *Energy cannot be created or destroyed, but only changed from one form to another.*
- *Most of what goes on in the universe–from exploding stars and biological growth to the operation of machines and motion of people– involves some form of energy being transformed into another. Energy in the form of heat is almost always one of the products of an energy transformation.*
- *Inspect, disassemble, and reassemble simple mechanical devices and describe what the various parts are for; estimating what the effect that making a change in one part of a system is likely to have on the system as a whole.*

Math
Measuring
Using formulae

Science
Physical science
　　simple machines
　　　　gears
　　compound machines
　　　　gear trains
　　conservation of energy

Integrated Processes
Observing
Collecting and recording data

Interpreting data
Comparing and contrasting
Generalizing

Materials
LEGO® elements (per group):
　1　building plate
　4　1 x 4 beams
　3　1 x 16 beams
　4　2 x 4 bricks
　2　bushings
　2　axles, 6-studs long
　1　gear, 8-tooth
　1　gear, 40-tooth
String
Metric ruler
Spring scale

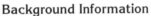

Background Information
　　Gears are a modification of the wheel and axle. As with all simple machines, gears increase/reduce a force or the distance over which a force is applied; or they may change the direction in which the force is applied.

　　Gears always work in teams. Any two or more gears working together is called a *gear train*. The gear on the train to which the force is initially applied is called the *driver*. The final gear on the train to which the force is transferred is called the *driven* gear. A gear train combines two wheel and axles, or two simple machines. This makes a gear train a compound machine.

　　Gears follow the *law of the conservation of energy*. Ideally the work done on the eight-tooth gear side is the same as the work done on the 40-tooth gear side. The eight-tooth gear side spreads a small force over a long distance, and the 40-tooth gear side uses a large force over a short distance.

　　The gear sizes allow us to predict how these forces and distances will compare. The eight-tooth gear and the 40-tooth gear are different in size by a factor of five. To pull on the eight-tooth gear's axle will require one-fifth the force of pulling on the 40-tooth gear's axle, but you will need to pull five times as much string on the eight-tooth side. Inversely, it takes five times the force to pull on the 40-tooth side, but you only have to pull it one-fifth the distance.

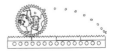

As the string on the eight-tooth gear is pulled, it will take very little force, but it will need to be pulled a long distance. When the string on the 40-tooth gear is pulled, it will take a great deal of force, but very little distance.

Management

1. This activity is best done in groups of two.
2. Before beginning the activity, build the model to use as a demonstrator.
3. The "magic" in this activity only appears during the demonstration in which you pull the strings. Once **MAGIC, continued from page** the students begin to assemble the model, they will understand that the different-sized gears require different lengths of string.

Procedure

1. Demonstrate the model to the students and discuss the *Key Question*.
2. Direct students to assemble the model as shown in the illustration. In assembling the model, the 15 cm string is attached to the 40-tooth gear's axle and the 40 cm string is attached to the eight-tooth gear's axle. Have the students extend as much of the 15 cm string as possible, but wind the 40 cm string around its axle.
3. After the model is constructed, invite students to first pull on the 40 cm string and then the 15 cm string, making qualitative observations of the distance pulled and force used. Have them record their observations.

4. Have students measure and record the distance the bricks move for each string.
5. Direct them to use the spring scale to measure the force used to pull each string. Have them record this information.
6. Discuss the results and the generalizations that can be made.

Discussion

1. Which string was easier to pull? [the eight-tooth gear's string]
2. Which string was harder to pull? [the string on the 40-tooth gear]
3. Which string did you pull a long distance? [the string on the eight-tooth gear]...a short distance? [the string on the 40-tooth gear]
4. Compare and contrast the force you used and the distances you pulled the strings? [They are opposite or inverse.]
5. How did this machine make this happen?
6. How could you use this machine?

Extensions

1. Have students use other gear combinations to see what happens.
2. Have students construct a machine using this gearing concept.

Magic String

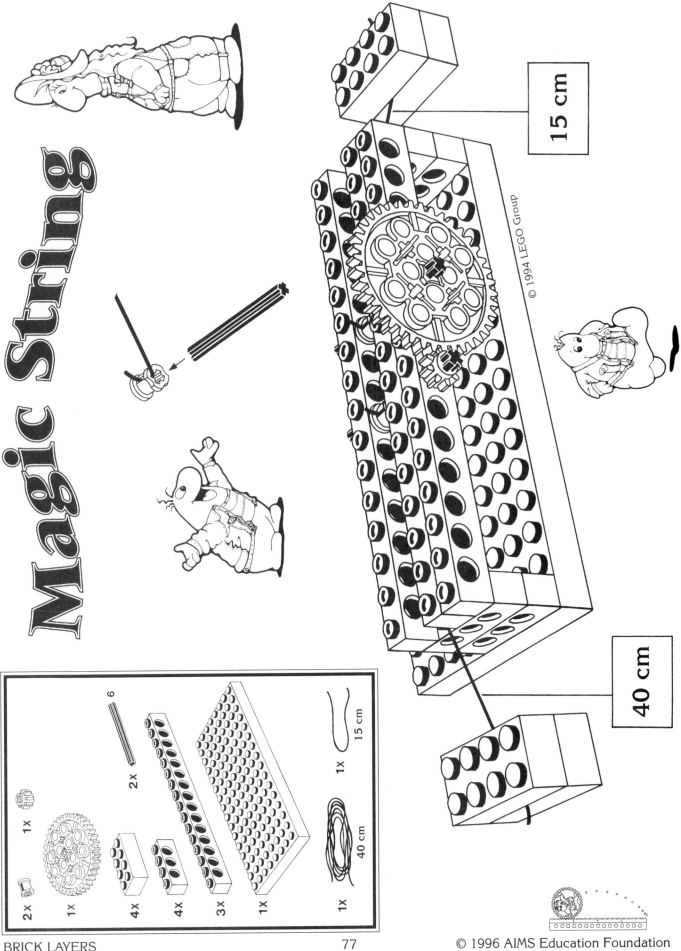

15 cm

40 cm

© 1994 LEGO Group

6

2x

2x

1x

4x

4x

3x

1x

1x

1x

15 cm

40 cm

77

Magic String

Build the model and see what happens when you pull the strings. Record your observations and measurements.

8-Tooth Gear

Distance of Pull	Force of Pull

40-Tooth Gear

Distance of Pull	Force of Pull

Explain the results.

Meshing Around: Exploring Gear Trains

Topic
Simple Machines: Wheels, Gears

Key Question
How do the sizes and positions of gears in a gear train affect how they move?

Focus
Students mesh different pairs of gears to see how their sizes affect their turning. This activity is designed to develop an intuitive understanding of gears that will be quantified and formalized in other activities.

Guiding Documents
NCTM Standards
- *Generalize solutions and strategies to new problem situations*
- *Discuss mathematical ideas and make conjectures and convincing arguments*
- *Validate their own thinking*

Project 2061 Benchmarks
- *Inspect, disassemble, and reassemble simple mechanical devices and describe what the various parts are for; estimating what the effect that making a change in one part of a system is likely to have on the system as a whole.*

Math
Identifying patterns

Science
Physical science
  simple machines
    wheel and axle
    compound machines
    gears

Integrated Processes
Observing
Collecting and recording data
Interpreting data
Comparing and contrasting
Generalizing

Materials
LEGO® elements (per group):
1   1 x 16 beam
4   axles, 6-studs long
2   axles, 8-studs long
2   gears, 40-tooth
3   gears, 24-tooth
4   gears, 8-tooth
6   bushings
Scissors
Glue

Background Information
Gears are a modification of the wheel and axle. As with all simple machines, gears increase/reduce a force or the distance over which a force is applied; or they may change the direction in which a force is applied.

Gears always work in teams. Two gears working together is a combination of two simple machines. This makes a pair of gears a *compound machine*.

Two or more gears working together is called a *gear train*. The gear on the train to which the force is initially applied is called the *driver*. The final gear on the train to which the force is transferred is called the *driven* gear. Any gears between the driver and driven gears are called *idlers*.

Gears in a train rotate at a rate relative to the driver's rotation. The rate of rotation of the gears is inverse to the relative size of the gears. If a gear is larger than the driver, it will rotate slower than the driver. If a gear is smaller than the driver, it will rotate faster than the driver.

The sequence position of a gear in a train determines its direction of rotation in relation to the driver. All gears in odd sequence positions (1st or driver, 3rd, 5th, etc.) will rotate in the same direction as the driver. All gears in even sequence positions (2nd, 4th, 6th, etc.) will rotate in the opposite direction of the driver.

Management
1. This activity is best done in groups of two.
2. Allow students time to explore and develop their own theories of what is going on with the gear trains. They will need this intuitive experience before moving to a more formal understanding.
3. Encourage dialogue between students, having them share with each other their insights, but emphasize that exploratory time is not merely social time.

Procedure
1. Distribute the materials to every pair of students.
2. Have them build a simple gear train (ex. 24:8:40).

3. Explain to students the terms of gear trains: *driver*, *driven gear*, and *idler*. This common language will help in the discussion of their findings.
4. Discuss the *Key Question* with the class.
5. Distribute the student record sheets, scissors, and glue. Explain to the students that they must keep a record of the gear trains they construct and any observations they make.
6. Allow students time to explore by building their own gear trains and making records of their observations.
7. Direct students to cut out the pictures of the gears they used in the gear trains and glue them onto the picture of the beam. Have them write or illustrate their observations under the gear trains.

Suggestions: If students need more guidance in making records of their observations, suggest some of the following:
- Make a list of all the combinations of two gears that can be made and try those.
- Put arrows to show direction of rotation or record as clockwise or counterclockwise.
- Record relative rates as slower, faster, or the same.
- Try more than two gears in a train.

7. End the activity with students discussing what they have discovered about gear trains.

Discussion
1. When you built a two-gear train, what did you notice about the direction of rotation of the driven gear and the driver? [opposite direction]

2. When you built a three-gear train, what did you notice about the direction of rotation of the driven gear and the driver? [same direction] What direction did the idler rotate? [opposite direction]
3. When you built a train with four or more gears, what did you notice about the direction of rotation of the driven gear and the driver? [opposite direction] What direction did the idlers rotate?
4. What generalization can you make about the directions the gears rotate? [odd positions rotate the same as driver, even positions opposite driver]
5. When a gear is the same size as the driver, how does its rate of rotation compare to the driver's rate? [same]
6. When a gear is larger than the driver, how does its rate of rotation compare to the driver's rate? [slower]
7. When a gear is smaller than the driver, how does its rate of rotation compare to the driver's rate? [faster]

Extension
Have students create and construct a LEGO® machine that contains a gear train.

Home Link
Have students identify items that use gears and make a list with the name of the item and where they saw it. If they can see the gears, determine if the train is making the driven gear go faster of slower than the driver.

80

Meshing Around: Exploring Gear Trains

Meshing Around: Exploring Gear Trains

Glue the gear trains you tried on the beams. Record your observations.

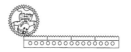

Reel Changes

Topic
Simple Machines: Gears

Key Question
What happens when you change the size of the gears on a fishing reel?

Focus
In this activity, students will construct a fishing reel using a single gear train. By changing the gears in the train, they will discover the gears' effect on the rate at which the line is reeled in.

Guiding Documents
NCTM Standards
- *Explore problems and describe results using graphical, numerical, physical, algebraic, and verbal mathematical models or representations*
- *Construct, read, and interpret tables, charts, and graphs*
- *Analyze tables and graphs to identify properties and relationships*

Project 2061 Benchmarks
- *Energy cannot be created or destroyed, but only changed from one form into another.*
- *Most of what goes on in the universe – from exploding stars and biological growth to the operation of machines and the motion of people – involves some form of energy being transformed into another. Energy in the form of heat is almost always one of the products of an energy transformation.*
- *Inspect, disassemble, and reassemble simple mechanical devices and describe what the various parts are for; estimating what the effect that making a change in one part of a system is likely to have on the system as a whole.*

Math
Graphing
Identifying patterns

Science
Physical science
    simple machines
    compound machines
    gears

Integrated Processes
Observing
Collecting and recording data
Interpreting data
Comparing and contrasting
Generalizing

Materials
LEGO® elements (per group):
1	weighted brick
3	2 x 6 plates
4	2 x 4 plates
4	1 x 16 beams
1	1 x 2 beams
2	axles, 6-studs long
1	connector peg
2	bushings
2	gears, 40-tooth
2	gears, 24-tooth
2	gears, 8-tooth
1	steering wheel

String, 60 cm
Scissors (optional)
Tape (optional)

Background Information
Gears are a modification of the wheel and axle. As with all simple machines, gears increase/reduce a force or the distance over which a force is applied; or they may change the direction in which a force is applied.

Gears always work in teams. Any two or more gears working together is called a *gear train*. The gear on the train to which the force is initially applied is called the *driver gear*. The final gear on the train to which the force is transferred is called the *driven gear*.

The number of teeth on the gears used in a train determines how the force and distance will change. If the driver gear has less teeth than the driven gear, the driven gear will rotate fewer times and at a slower rate, but with more force than the driver gear. Inversely, if the driver gear is larger than the driven gear, the driven gear will rotate more times and at a faster rate, but with less force than the driver gear.

Consider the LEGO® model in which the gear on the crank (driver gear) is a 24-tooth gear and the gear on the reel (driven gear) is an eight-tooth gear. Every time the crank is rotated, the gears rotate 24 teeth. Twenty-four teeth is one complete rotation of the driver gear, but three of the driven gear. For each rotation of the driver gear, the driven gear rotates three turns. The driven gear is moving at three times the rate.

Gears, as with all simple machines, follow the laws

of the *conservation of energy*. As force moves a greater distance, it does so with less force. As the driven gear rotates at three times the rate of the driver gear, it also develops a force which is one-third as great.

Suppose the fishing reel axle has to be rotated 30 times to wind in the weighted brick. Return to the previous example where the driven gear turns at three times the rate but with one-third the force of the driver gear. The student must turn the crank one time to rotate the reel three times. It will take ten cranks to reel in the weight. With the force being one-third as great on the driven gear, the students will use three times the force and it will seem difficult to lift the weight.

Management
1. This activity is best done in groups of two.
2. Warn students not to compress the beams with the gears, bushings, or crank. Compressing will not allow the axles to turn freely and students will not be able to feel the differences in the gear trains.
3. The model requires that students change the axle positions. The gear trains on the student page have been arranged to minimize the axle changes.

Procedure
1. Discuss the *Key Question*.
2. Give students time to assemble the rod and reel as shown in the illustration. The fish may be cut out and taped to the weighted brick.
3. Instruct students to reel in the weighted brick (the fish) counting how many turns they make on the crank to lift it all the way.
4. Have them record how many turns they counted and make a bar graph of the data.
5. Direct students to change the crank gear and reel axle gear to a different arrangement shown on the chart, and repeat steps four and five. It may be necessary to point out that the axles may need to be moved to accomodate the different gear arrangements.
6. When all nine arrangements of gear trains have been tested, urge students to determine a pattern to the gear train arrangement and the number of cranks required to raise the weight.

Optional
7. Have students cut to separate the nine bars of the graph and arrange them in order from most turns to least turns to help determine how the gear trains affect the rate of rise.

Discussion
1. What arrangements took the same amount of turns to lift the weight? [8:8, 24:24, 40:40]
2. What do these arrangements have in common? [Both gears are the same size.]
3. How do these arrangements differ? [Each arrangement used different sized gears.]

4. What arrangements took the least turns to lift the weight? [24:8, 40:8, 40:24]
5. What do these arrangements have in common? [The crank gear (driver) is bigger than reel axle gear (driven).]
6. How do these arrangements differ? [different sized gears]
7. What arrangements took the most turns to lift the weight? [8:24, 8:40, 24:40]
8. What do these arrangements have in common? [The crank gear (driver) is smaller than reel axle gear (driven).]
9. How do these arrangements differ? [different sized gears]
10. When was it easiest to lift the weight with the crank? [When the driver gear is smaller.]
11. When was it hardest to lift the weight with the crank? [When the driver gear is larger.]
12. Generalize what you have learned about how to arrange gears.
13. To reel in a whale, what type of gear arrangement would you want? [little driver: big driven]
14. To reel in a small fish to keep it away from a pursuing predator, what type of gear arrangement would you want? [big driver: small driven]

Extension
Bring a fishing reel into the classroom and have students determine the gear ratio used.

Reel Changes

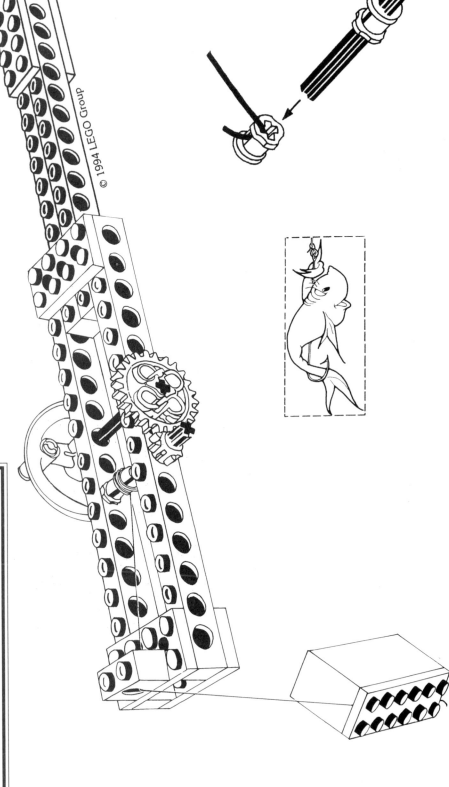

© 1994 LEGO Group

Parts Inventory

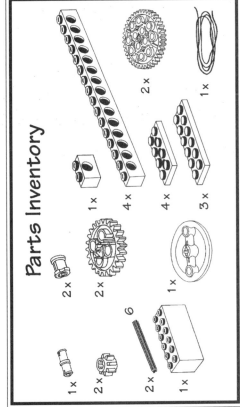

2 x

1 x

4 x

4 x

3 x

2 x

2 x

1 x

2 x

6

1 x

2 x

1 x

BRICK LAYERS

85

Reel Changes

How many turns does it take to reel in the "fish?"

	Turns	Gears
		40 / 8

| 70 | 60 | 50 | 40 | 30 | 20 | 10 | 0 |

	Turns	Gears
		40 / 40

| 70 | 60 | 50 | 40 | 30 | 20 | 10 | 0 |

	Turns	Gears
		40 / 24

| 70 | 60 | 50 | 40 | 30 | 20 | 10 | 0 |

	Turns	Gears
		24 / 40

| 70 | 60 | 50 | 40 | 30 | 20 | 10 | 0 |

	Turns	Gears
		24 / 8

| 70 | 60 | 50 | 40 | 30 | 20 | 10 | 0 |

	Turns	Gears
		24 / 24

| 70 | 60 | 50 | 40 | 30 | 20 | 10 | 0 |

	Turns	Gears
		8 / 8

| 70 | 60 | 50 | 40 | 30 | 20 | 10 | 0 |

	Turns	Gears
		8 / 40

| 70 | 60 | 50 | 40 | 30 | 20 | 10 | 0 |

	Turns	Gears
		8 / 24

| 70 | 60 | 50 | 40 | 30 | 20 | 10 | 0 |

Driven Gear
Driver Gear

BRICK LAYERS

86

Turn Around

Topic
Simple Machines: Gears

Key Question
If you know the gear sizes in a gear train and the number of turns the driver gear makes, how can you determine the number of turns the driven gear will make?

Focus
Students pair different gear combinations. By counting the number of rotations each gear makes, they discover the inverse relationship of gear size to rate of rotation. They also note the effort of cranking the different gear trains to recognize the trade-off made between distance and force.

Guiding Documents
NCTM Standards
- *Generalize solutions and strategies to new problem situations*
- *Construct, read, and interpret tables, charts and graphs*
- *Analyze tables and graphs to identify properties and relationships*

Project 2061 Benchmarks
- *Energy cannot be created or destroyed, but only changed from one form into another.*
- *Inspect, disassemble, and reassemble simple mechanical devices and describe what the various parts are for; estimate what the effect that making a change in one part of a system is likely to have on the system as a whole.*

Math
Identifying patterns
Using computation
Using and applying formulae

Science
Physical science
   conservation of energy
   simple machines
   compound machines
     gears

Integrated Processes
Observing
Collecting and recording data
Comparing and contrasting
Generalizing
Predicting

Materials
LEGO® elements (per group):
  4   2 x 4 bricks
  2   1 x 8 beams
  1   2 x 6 plate
  2   axles, 6-studs long
  2   connector pegs
  1   steering wheel
  2   bushings
  1   piston rod
  2   gears, 40-tooth
  2   gears, 24-tooth
  2   gears, 8-tooth
  1   beveled gear
  1   building plate
  1   weighted brick
String, 35 cm

Background Information
Gears are a modification of the wheel and axle. As with all simple machines, gears increase/reduce a force or the distance over which a force is applied; or they change the direction in which a force is applied.

Gears always work in teams. Any two or more gears working together is called a *gear train*. The gear on the train to which the force is initially applied is called the *driver gear* (input). The final gear on the train to which the force is transferred is called the *driven gear* (output).

By multiplying the number of teeth on the input gear (T_i) by the number of rotations it makes (R_i), the number of teeth moved by the input gear can be determined. When two gears mesh, the same number of teeth must move on both gears; the teeth turned on the output gear is equal to the number of teeth turned by the input gear. Like the input gear, the teeth moved on the output is a product of the number of teeth on the gear (T_O) and the rotations made by that gear (R_O). This equal relationship can be written as the equation: $T_i \times R_i = T_O \times R_O$.

This relationship allows you to determine the rotations of the driven gear if you know the sizes and number of rotations on the driver gear. Consider the example of a 24-tooth gear meshed with a 40-tooth gear. If you rotate the 24-tooth gear five times, that will cause 120 teeth to move (24 x 5 = 120). One hundred and twenty teeth on the 40-tooth gear must move as a result. To do this, the 40-tooth gear will need to turn three times (120 ÷ 40 = 3).

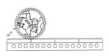

With the 24:40 gear example, the driver gear turns five times for three turns on the driven. The driver gear's axle is turning faster and going farther around than the driven gear's axle. Since the gears are meshed, they will do the same amount of work. The smaller gear does the work by going farther with less force. The larger gear does the work by going a smaller distance but with more force. This follows the law of the conservation of energy. The gears are making a trade-off, more distance with less force. The 24:40 gear example allowed the weight to be raised by making more turns with the driver gear, but doing it with less force.

Management

1. This activity requires at least two students per group; one turns the driver gear and counts the number of rotations on the crank, while the second counts the number of rotations on the driven gear.
2. When the larger gear is the driver gear, the driven gear will turn more rapidly. Caution the students to turn the crank slowly in this situation to allow the second person to count the rotations.
3. Emphasize the importance of accurate record keeping.
4. The two blank columns on the chart are provided as helping columns. After gathering the data, students need time to analyze them to see if they can find any patterns. The blank columns are provided so students can experiment and write the solutions to different operations they do to the numbers on each side. They should discover that the products of the numbers on each side form an equality. These products can be recorded in the columns.

Procedure

1. Discuss the *Key Question*.
2. Distribute a set of LEGO® materials to every two students.
3. Allow students time to build the model.
4. Have students construct the various gear trains listed on the chart. Tell them that while one student turns the driver crank the number of turns listed on the chart, the other student counts the turns on the driven gear. Direct the students to record the turns of the driven gear. As students construct the various gear trains, advise them that they will need to reposition the axles to accomodate the gears.
5. Allow students time to determine the relationship between the driver and driven gear numbers.
6. Refer again to the *Key Question*. Have students discuss how the relationships they found could be used to answer the question.
7. Have students select a gear train and the number of driver turns and predict the number of driven turns. Have them verify their predictions with the gear train.

Discussion

1. Which gear train gave the most output turns for each input turn? [40:8]
2. Which gear train would lift the brick the fastest? [40:8]
3. With which gear train was it hardest to lift the brick? [40:8]
4. Which gear train gave the least output turns for each input turn? [8:40]
5. Which gear train would lift the brick the slowest? [8:40]
6. With which gear train was it easiest to lift the brick? [8:40]
7. Which gear trains gave the same output turns as were put in? [8:8, 24:24, 40:40]
8. What relationship do you see between the numbers on the driver side of the chart and those on the driven side? [When the two numbers on each side are multiplied, their products are the same.]
9. How could you write your pattern as an equation? [$T_i \times R_i = T_o \times R_o$]
10. If a 16-tooth gear were meshed with a 40-tooth gear, what do you predict would happen in turns, rate, and effort? [two turns out for five turns, slower, relatively easy]
11. What are the advantages and disadvantages of a gear train with a small driver and large driven gear? [easier to lift things, more turns or slower lift]
12. What are the advantages and disadvantages of a gear train with a large driver and small driven gear? [less turns or faster lift, harder to lift things]
13. What application can you think of when you would want a small driver gear and a large driven gear? Why?
14. What application can you think of when you would want a large driver gear and a small driven gear? Why?

Extension

Find pictures of actual winches (LEGO® Activity Card Pack 1031, card #6) and explain what the winch is doing as far as effort and rate.

88

Turn Around

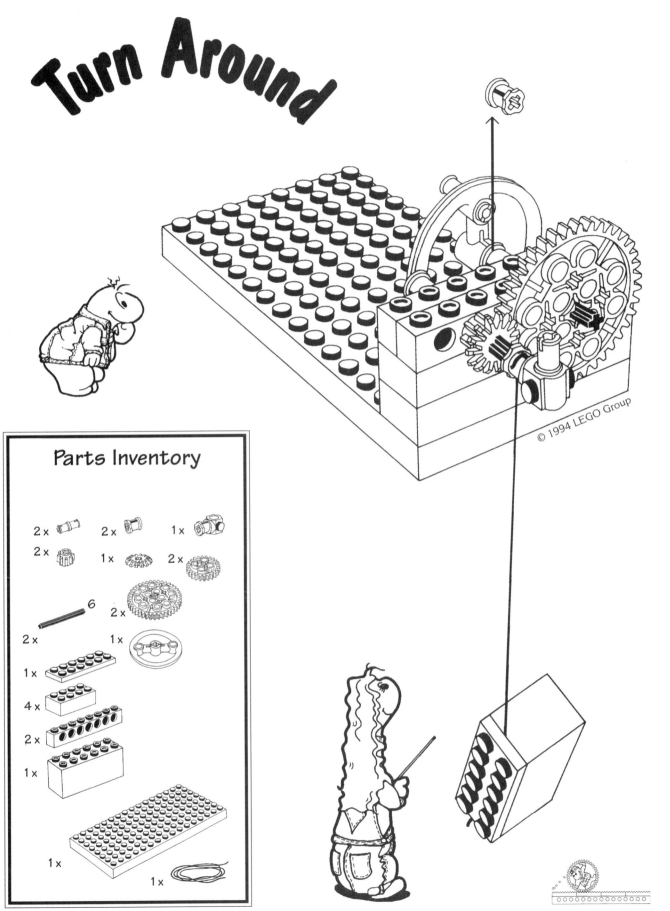

Parts Inventory

2 x   2 x   1 x
2 x   1 x   2 x
6
2 x   1 x
2 x
1 x
4 x
2 x
1 x
1 x
1 x

© 1994 LEGO Group

Turn Around

How many turns does the driven gear have to make?

© 1994 LEGO Group

Driver				Driven	
Teeth	Turns			Teeth	Turns
40	1			8	
40	2			8	
40	3			24	
40	6			24	
40	9			24	
40	1			40	
24	1			8	
24	2			8	
24	3			8	
24	1			24	
24	5			40	
24	10			40	
8	1			8	
8	3			24	
8	6			24	
8	9			24	
8	5			40	
8	10			40	

Use these columns to show your work.

Dial-A-Gear

Topic
Simple Machines: Gears

Key Question
If you know the size of the gears and how much the driver gear is rotated, how can you predict how much a driven gear will turn?

Focus
Students pair different gear combinations. By counting the number of rotations the driven gear turns for each rotation of the driver gear, they find the gear ratio. They see this relationship represented as a ratio, on a graph, and as an equation.

Guiding Documents
NCTM Standards
* *Explore problems and describe results using graphical, numerical, physical, algebraic forms*
* *Develop number sense for whole numbers, fractions, decimals, integers, and rational numbers*
* *Understand and apply ratios, proportions, and percents in a wide variety of situations*
* *Represent numerical relationships in one- and two-dimensional graphs*

Project 2061 Benchmarks
* *Energy cannot be created or destroyed, but only changed from one form into another.*
* *Mathematical statements can be used to describe how one quantity changes when another changes. Rates of change can be computed from magnitudes and vice versa.*
* *The graphic display of numbers may help to show patterns such as trends, varying rates of change, gaps, or clusters. Such patterns sometimes can be used to make predictions about the phenomena being graphed.*

Math
Using rational numbers
    equivalent fractions
Graphing
Identifying patterns
Using and applying formulae

Science
Physical science
    simple machines
    compound machines
        gear ratios

Integrated Processes
Observing
Collecting and recording data
Interpreting data
Comparing and contrasting
Generalizing

Materials
LEGO® elements (per group):
  4 1 x 16 beams
  2 2 x 8 plates
  2 2 x 4 plates
  3 bushings
  1 connector peg
  2 axle, 6-studs long
  2 pulley wheels
  1 steering wheel
  2 gears, 8-tooth
  2 gears, 24-tooth
  2 gears, 40-tooth
Tagboard
Scissors

Background Information
Gears are a modification of the wheel and axle. As with all simple machines, gears increase/reduce a force or the distance over which a force is applied; or they may change the direction in which a force is applied.

Gears always work in teams. Any two or more gears working together is called a *gear train*. The gear on the train to which the force is initially applied is called the *driver gear*. The final gear on the train to which the force is transferred is called the *driven gear*.

The ratio of teeth on the gears has a interesting relationship to both the turning ratio of the gears, and the ratio of forces required to turn the gears.

Consider the example of a gear train with an eight-tooth gear as the driver gear and a 40-tooth gear as the driven gear. The driver gear is one-fifth the size of the driven gear (8/40=1/5). Inversely, the driven gear is five times as large as the driver gear. When the gears are rotated, the driver gear rotates five times for the driven gear's one (8 x 5 = 40 x 1). The turning ratio of the driver to driven gear is five to one (5/1). You could say the driver gear rotates at five times the rate of the driven gear, or inversely the driven gear rotates at one-fifth (1/5) the rate of the driver gear. Following the law of the conservation of energy, the force will change inversely to the change in distance. Since the driven gear

is turning at one-fifth (1/5) the rate of the driver gear, it will apply five (5/1) times the force to its axle that is being applied to driver gear's axle. This change in force is called the *mechanical advantage*.

These relationships allow an engineer to choose the correct gears for a specific application. The turning ratio and mechanical advantage of the gear train can be found by making a ratio comparing the teeth on a driver gear to the teeth on the driven gear. If a 40-tooth gear is meshed with a 24-tooth gear, the tooth ratio is 40 to 24 (40/24) which is simplified to five-thirds (5/3 = 1 2/3). This ratio can be applied to the turning ratio. The driven gear turns at one and two-thirds the rate of the driver gear. The driven gear is turning at a faster rate than the driver gear. Inversely, it is turning with less force. To calculate this force, take the reciprocal of the tooth ratio, which in this case is three-fifths(3/5). The mechanical advantage is three-fifths. This might better be called a mechanical disadvantage since this machine gives you less force than that with which you started.

Connecting an eight-tooth driver gear with a 40-tooth driven gear will have the inverse effect of meshing a 40-tooth driver gear with an eight-tooth driven gear. When this inverse relationship is represented on a graph, this relationship takes on new meaning. When the turning rates of the six available inverse combinations of gears are graphed, a symmetrical pattern evolves. The six lines have slopes of 1/5, 1/3, 3/5, 5/3, 3/1, 5/1. The line of symmetry for the pattern is the line with a slope of one. This graphic display helps students to realize that the order of gears in a train has an inverse effect in the train's functioning.

The activity is enriched as students recognize that the gears' turning rates can be represented by ratios or linear graphs, and these can be used to write equations.

Recording gear ratios in fraction form allows students to discover the relationship of teeth to turns. Gear ratios are conventionally recorded with decimal notation. It would be valuable for students to recognize that the fractional form of gear ratios can be written in decimal notation.

Management

1. This activity requires at least two people in a group. One person turns the driver gear and counts the number of rotations on the crank. The second person counts the number of rotations on the driven gear.

2. When the larger gear is the driver gear, the driven gear will turn more rapidly. Caution the students to turn the crank slowly in this situation so the second person can count the driven gear's rotations.

3. Before class, copy the counting wheels onto tagboard. If copying onto tagboard is not possible, copy the counting wheels on paper. Have scissors and glue available so students can cut out the paper counting wheel and glue it to the tagboard.

4. Before and during the activity, emphasize that students need to record the number of complete rotations along with the remaining fractional part of a rotation.

5. The *Procedure* contains two parts. *Part 1* develops the understanding of the relationship of number of teeth and the number of rotations. *Part 2* is included for students who are able to use counting wheels to change fractions into equivalent decimal forms.

Procedure

Part 1

1. Distribute the LEGO® elements and have students construct the gear assembly as shown in the illustration.

2. Discuss the *Key Question* while referring to the assembly. Make sure the students understand that the driver gear is the one on the same axle as the steering wheel, and the driven gear is on the axle with the counting wheel.

3. Using the illustrated 8/24 combination, direct students to hold the assembly upright and align the steering wheel spoke with the beam. Keeping the steering wheel in this alignment, have students turn the counting wheel so the start position is aligned with the top of the beam. Urge them to make sure the *Fraction Counting Wheel* is snug between the two pulley wheels.

4. Direct the students to rotate the steering wheel and the attached driver gear one time clockwise. Have them record how far the *Fraction Counting Wheel* has rotated and use its simplest form. If necessary, inform them that this measurement is found on the counting wheel at the end of the beam.

5. Tell students to continue to rotate the driver gear, recording the rotations of the driven gear until the chart is completed. Students need to be aware that as the driven gear makes the second rotation, a mixed number will need to be recorded (1 5/15 = 1 1/3).

6. Allow time for students to complete all six charts with the different gear combinations. They will discover that the gear combinations of 24/8, 40/8, and 40/24 will increase the rate of the driven gear. Encourage them to move the drive gear very slowly so they can determine the total number of rotations.

7. Have students discuss what patterns they see in the charts.

8. Direct them to make line graphs of their data. Have them discuss what patterns they see in the graph and discuss why those patterns occurred.

9. Have students write an equation relating the driven rotation to driver rotations for each gear combination.

Part 2

10. Ask students to write their equations in decimal form. If they cannot, distribute the *Decimal Counting Wheel* and record sheet.

11. Ask the students to determine another way they could record 4/5, or 12/15, of a rotation. [0.8]

12. When students see how to use the counting wheels to make the conversions, have them determine the decimal equivalents. They may need some guidance in determining decimal equivalents ($0.06 < 1/15 < 0.07$).

13. Encourage students to use their calculators to convert the fractions to their decimal equivalents to check the accuracy of the counting wheels.

14. Have students convert their equations into decimal form.

Discussion

1. What patterns do you see in your charts? [Every driver rotation increased the driven rotation by the same amount. This increase is the simplified form of the gear combination's tooth ratio.]

2. What patterns do you see in your graph? Explain why they occur. [Straight lines— consistent change in driven gear; Steeper lines—bigger driver gear, more driven rotations for each driver rotation; Symmetrical around slope of 1; Gear combinations inverted; Turn ratio inverted]

3. Write an equation that tells you how many driven rotations you get if you know the number of driver rotations and the teeth on the gears. [driven rotations = (driver teeth/driven teeth) x driver rotations]

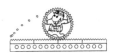

Dial-A-Gear

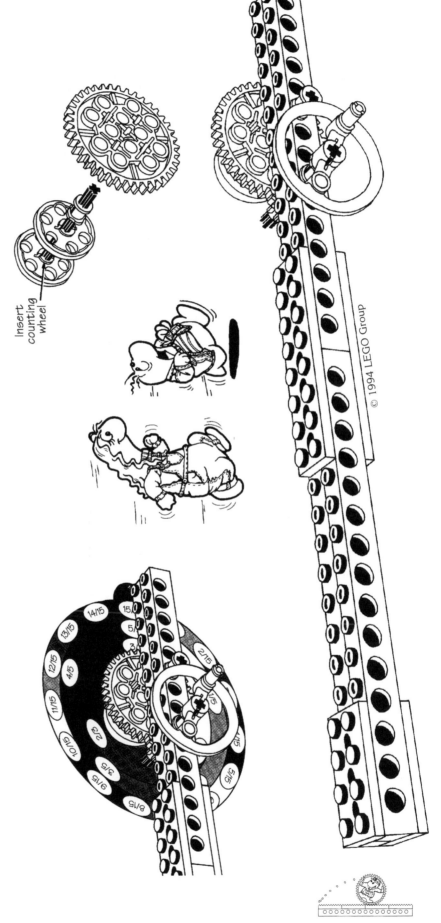

Insert counting wheel

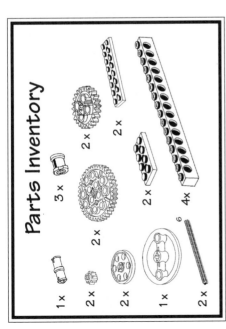

Parts Inventory

1 x
3 x
2 x

2 x
2 x
2 x

2 x

2 x

6

1 x

4 x

2 x

Dial-A-Gear

Fraction Counting Wheels

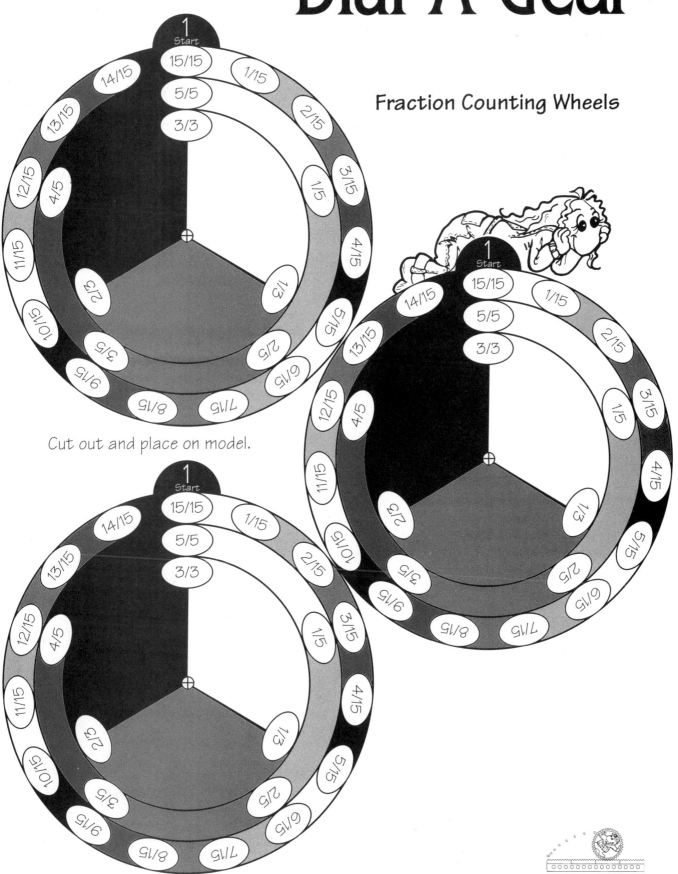

Cut out and place on model.

Dial-A-Gear

Teeth on Gears		
Driver /	Driven	= Simplified
8 /	24	=

Driver Rotations	Driven Rotations	Simplest Form
1		=
2		=
3		=
4		=
5		=

Teeth on Gears		
Driver /	Driven	= Simplified
8 /	40	=

Driver Rotations	Driven Rotations	Simplest Form
1		=
2		=
3		=
4		=
5		=

Teeth on Gears		
Driver /	Driven	= Simplified
24 /	40	=

Driver Rotations	Driven Rotations	Simplest Form
1		=
2		=
3		=
4		=
5		=

Teeth on Gears		
Driver /	Driven	= Simplified
24 /	8	=

Driver Rotations	Driven Rotations	Simplest Form
1		=
2		=
3		=
4		=
5		=

Teeth on Gears		
Driver /	Driven	= Simplified
40 /	8	=

Driver Rotations	Driven Rotations	Simplest Form
1		=
2		=
3		=
4		=
5		=

Teeth on Gears		
Driver /	Driven	= Simplified
40 /	24	=

Driver Rotations	Driven Rotations	Simplest Form
1		=
2		=
3		=
4		=
5		=

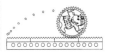

Dial-A-Gear

Use your data from your record sheet to make a broken-line graph for each gear train. You will have six different lines. Make each one a different color and label each one.

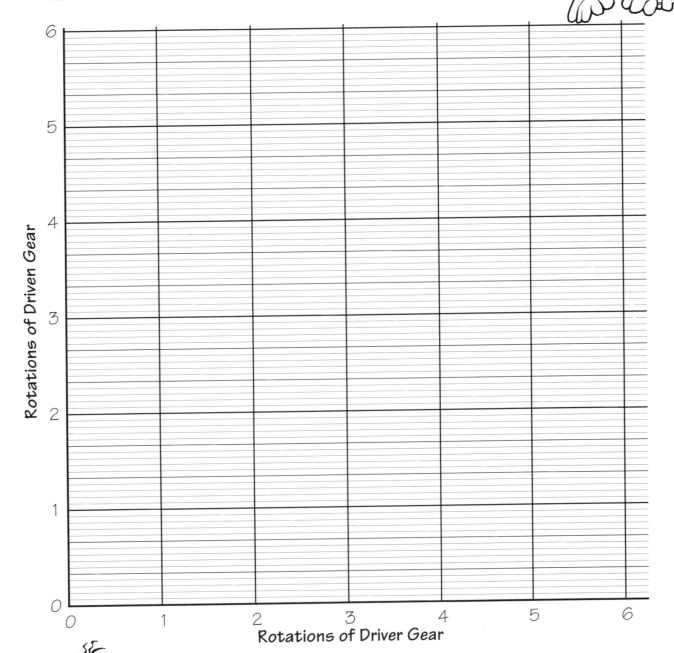

Rotations of Driven Gear (y-axis, 0 to 6)

Rotations of Driver Gear (x-axis, 0 to 6)

What patterns do you see between the numbers that describe each gear train and the line that resulted from rotating the driver gear?

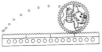

Dial-A-Gear

Decimal Counting Wheel

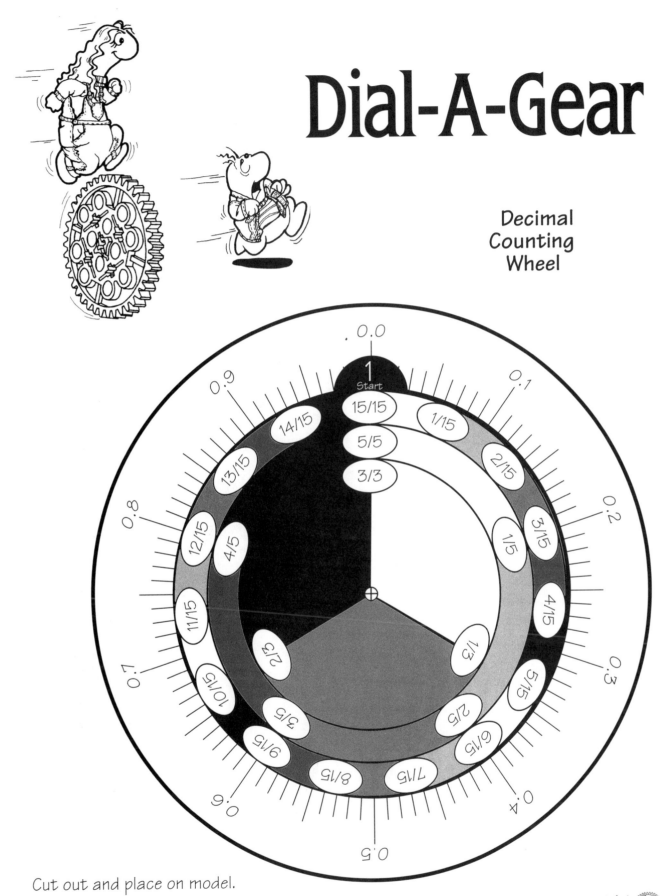

Cut out and place on model.

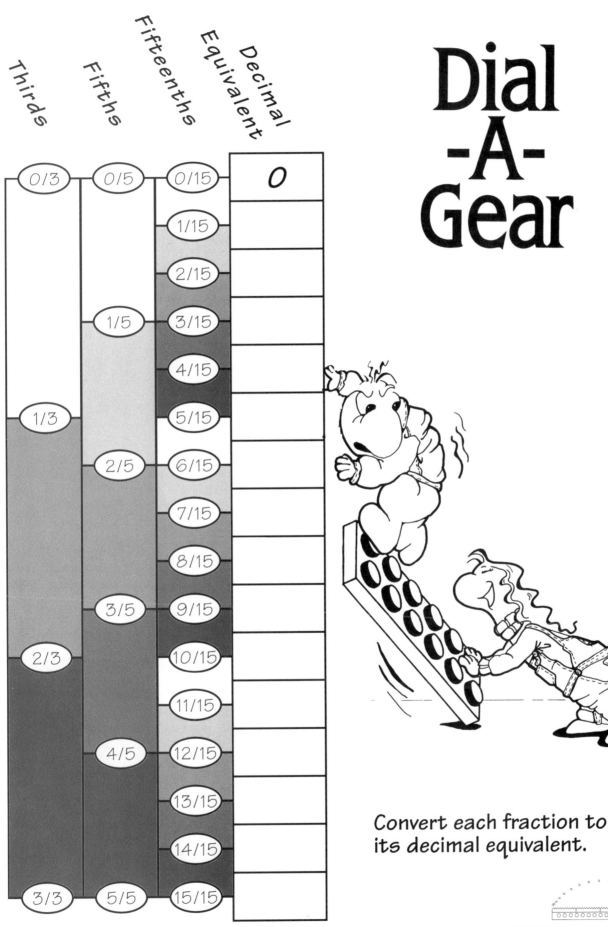

Dial -A- Gear

Convert each fraction to its decimal equivalent.

Thirds	Fifths	Fifteenths	Decimal Equivalent
0/3	0/5	0/15	0
		1/15	
		2/15	
	1/5	3/15	
		4/15	
1/3		5/15	
	2/5	6/15	
		7/15	
		8/15	
	3/5	9/15	
2/3		10/15	
		11/15	
	4/5	12/15	
		13/15	
		14/15	
3/3	5/5	15/15	

The Governor Rules

Topic
Simple Machines: Gears

Key Question
What gear combination works best to lift the governor to its highest position?

Focus
By changing the combination of gears, students will determine which gear train makes a governor turn fastest, and which train is easiest to turn. From this experience, students will come to recognize the trade-off made by the simple machine between distance and effort.

Guiding Documents
NCTM Standards
- *Analyze tables and graphs to identify properties and relationships*
- *Formulate problems from situations within and outside mathematics*
- *Generalize solutions and strategies to new problem situations*

Project 2061 Benchmarks
- *Energy cannot be created or destroyed, but only changed from one form to another.*
- *Most of what goes on in the universe–from exploding stars and biological growth to the operation of machines and motion of people–involves some form of energy being transformed into another. Energy in the form of heat is almost always one of the products of an energy transformation.*
- *Inspect, disassemble, and reassemble simple mechanical devices and describe what the various parts are for; estimate what the effect that making a change in one part of a system is likely to have on the system as a whole.*

Math
Identifying patterns

Science
Physical science
   simple machines
     gears
     conservation of energy
     centrifugal force

Integrated Processes
Observing
Collecting and recording data
Interpreting data
Comparing and contrasting
Generalizing

Materials
LEGO® elements (per group):

1	building plate
2	axles, 12-studs long
3	axles, 6-studs long
2	axles, 4-studs long
4	wheels with tires
1	steering wheel
2	gears, 8-tooth
1	gear, 40-tooth
1	gear, 24-tooth
1	beveled gear
1	crown gear
2	1 x 12 beams
2	1 x 8 beams
6	1 x 4 beams
8	1 x 2 beams
2	2 x 8 plates
4	2 x 4 plates
4	1 x 4 plates
4	1 x 3 plates
7	connector pegs
5	bushings
4	piston rods
1	pulley

Scissors
Glue

Background Information
Gears are a modification of the wheel and axle. As with all simple machines, gears increase/reduce a force or the distance over which a force is applied; or they may change the direction in which the force is applied.

Gears always work in teams. Any two or more gears working together is called a *gear train*. The gear on the train to which the force is initially applied is called the *driver*. The final gear on the train to which the force is transferred is called the *driven* gear.

The number of teeth on the gears used in a train determine how the force and distance will change. If the

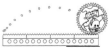

driver has less teeth than the driven gear, the driven gear will rotate less, but with more force than the driver. Inversely, if the driver is larger than the driven gear, the driven gear will rotate faster but with less force than the driver.

Consider the LEGO® model when the gear with the crank (driver) is a 40-tooth gear and the gear on the axle to the machine (driven gear) is an eight-tooth gear. Every time the crank is rotated one complete revolution, the gear rotates 40 teeth. Forty teeth is one complete rotation of the driver to five rotations of the driven gear (eight teeth times five rotations equals 40). For each rotation of the driver, the driven gear rotates five times. The driven gear is moving at five times the rate.

Gears, as simple machines, follow the *law of the conservation of energy*. As force moves a greater distance, it does so with less force. As the driven gear rotates at five times the rate of the driver, it also develops one-fifth the force but with faster rotational rate.

Centrifugal force is the force that propels the tires on the model outward from the center of rotation to the wide-opened position. In order to lift the governor to the highest position, the gear train must rotate the axle of the governor at a high rotational speed.

A governor is placed on an engine to control its speed. The engine is connected to the governor by an axle just as the model is connected to the crank. A valve is connected to the plate that spins around the center axis that rises as the governor rotates faster. As the engine speeds up, this plate rises, lifting the valve which decreases the amount of steam or fuel reaching the engine. As the speed decreases, the plate drops and opens the valve. In this way the speed of the engine is governed.

Management
1. This activity is best done in groups of two.
2. The model in this activity is relatively complicated. Allow an hour for students to assemble it. Allow 45 minutes to an hour for students to experiment with different gear combinations and discuss their conclusions.

Procedure
1. Discuss the *Key Question*.
2. Direct students to assemble the governor as shown in the illustration.
3. Have them cut out all the illustrations of gear train combinations.
4. Urge them to try all the illustrated gear trains, turning the crank and noting how fast the governor turns and how hard it is to turn the crank. Have them place the illustrations of the trains on the two continuums, sorting them as they try the different trains.

5. When students in a group agree on the arrangement of gear trains on the continuums, have them glue them down.
6. Have students discuss what conclusions they can draw from their investigations.

Discussion
1. What patterns do you see in the gear trains as they make the governor turn faster? [Largest gear to smallest gear is fastest. Smallest gear to largest gear is slowest.]
2. What patterns do you see in the gear trains as they become harder to turn? [Largest gear to smallest gear is hardest. Smallest gear to largest gear is easiest.]
3. Generalize what you have learned about how to arrange gears. [Fastest turning arrangement is also hardest to turn. Slowest turning arrangement is easiest to turn. The more a gear train increases the rate of rotation, the harder it is to turn.]
4. What type of gear arrangements would you want to use with a dizzying spin-type fair ride? [large driver gear with a small driven gear]
5. What type of gear arrangements would you want to use with a slow, peaceful merry-go-round ride? [small driver, large driven gear]

Extensions
1. Describe and/or design fair-type rides that use the principle of centrifugal force as the basis for operation.
2. Have students experiment with the placement of the wheels and tires on the governor's arms to gain a better understanding of leverage.

101

The Governor Rules

Parts Inventory

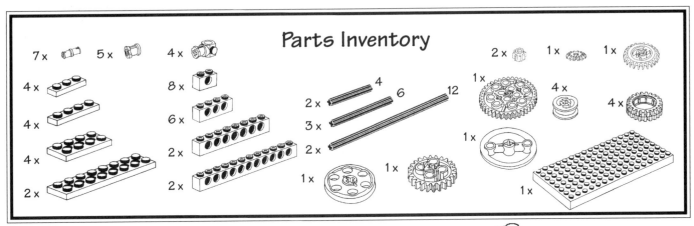

7 x	5 x	4 x
4 x	8 x	
4 x	6 x	2 x 4
4 x	2 x	3 x 6
2 x	2 x	2 x 12

2 x   1 x   1 x

1 x   4 x   4 x

1 x

1 x   1 x   1 x

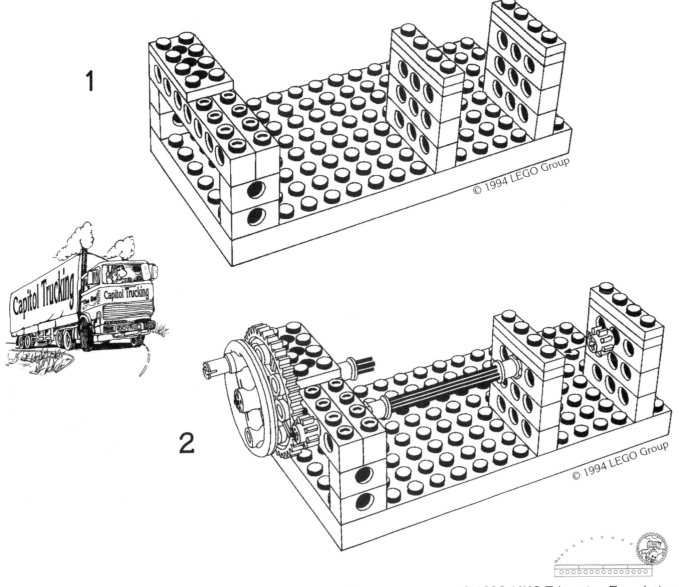

1

© 1994 LEGO Group

2

© 1994 LEGO Group

Capitol Trucking

102

The Governor Rules

3

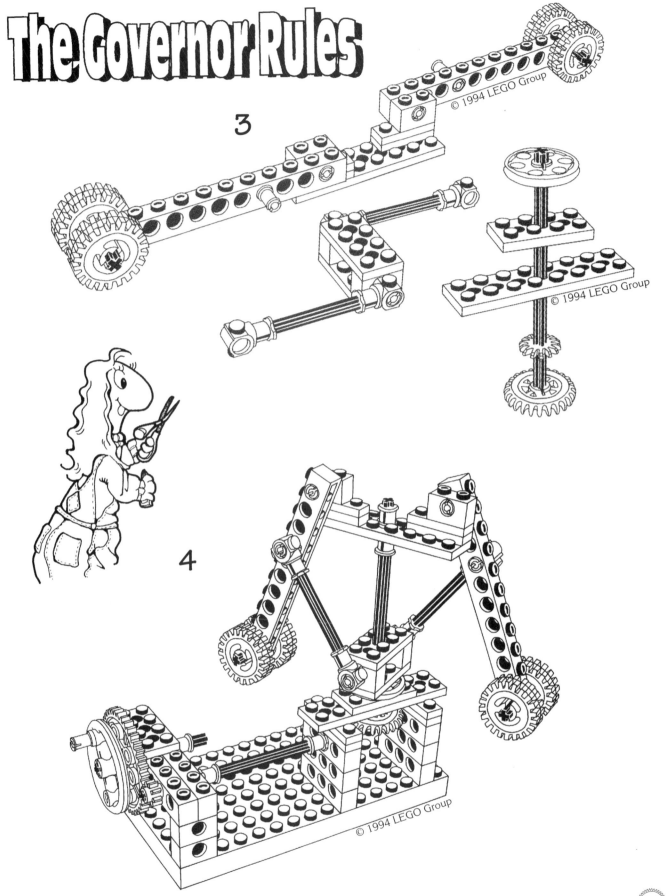

© 1994 LEGO Group

4

The Governor Rules

1. Cut out each of the six gear trains you can use for the governor.

2. Try each gear train on the governor. Observe how fast the governor turns and how difficult it is to turn.

3. On the next page, put the pictures in order of how fast and easy each gear train makes the governor turn.

Gears for Turning Rate

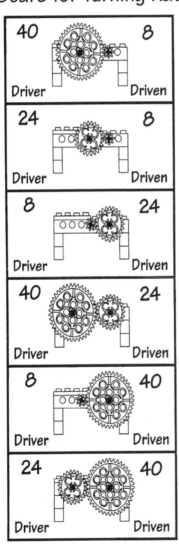

Gears for Ease of Turn

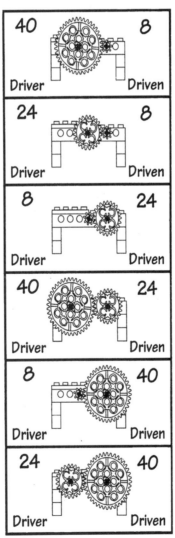

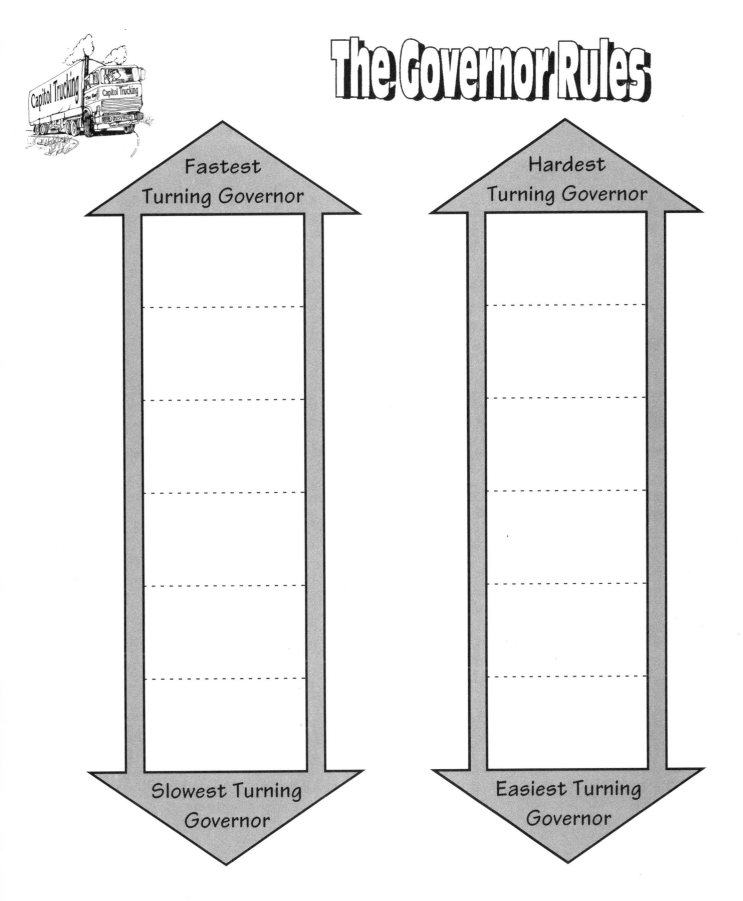

The Governor Rules

Fastest Turning Governor

Slowest Turning Governor

Hardest Turning Governor

Easiest Turning Governor

What do you notice about how fast a gear train makes the governor turn and how hard it is to turn the governor?

Gear Guessing

Topic
Simple Machines: Gears

Key Question
How can you determine what is inside this machine without taking it apart?

Focus
In this activity students apply what they know about gears to replicate a machine made with LEGO® elements with which they can only determine the input and output.

Guiding Documents
NCTM Standards
- *Formulate problems from situations within and outside mathematics*
- *Develop and apply a variety of strategies to solve problems, with emphasis on multistep and nonroutine problems*
- *Understand and apply reasoning processes, with special attention to spatial reasoning and reasoning with proportions and graphs*
- *Apply mathematical thinking and modeling to solve problems that arise such as art, music, psychology, science, and business*

Project 2061 Benchmarks
- *Scientists differ greatly in what phenomena they study and how they go about their work. Although there is not a fixed set of steps that all scientists follow, scientific investigations usually involve the collection of relevant evidence, the use of logical reasoning, and the application of imagination in devising hypotheses and explanations to make sense of the collected evidence.*
- *Thinking about things as systems means looking for how every part relates to others. The output from one part of a system (which can include material, energy, or information) can become the input to other parts. Such feedback can serve to control what goes on in the system as a whole.*

Math
Rotational measurement
Proportional reasoning
Identifying patterns
Logical reasoning

Science
Physical science
 simple machines
 compound machines
 gear ratios

Integrated Processes
Observing
Collecting and recording data
Predicting
Inferring
Making and testing hypotheses
Applying

Materials
LEGO® elements (per machine):

1	building plate
4	1 x 16 beams
8	1 x 4 beams
4	1 x 2 beams
2	2 x 4 plates
2	axles, 6-studs long
3	axles, 4-studs long
2	connector pegs
2	gears, 40-tooth
2	gears, 24-tooth
3	gears, 8-tooth
2	piston rods
5	bushings

Tape
Scissors

Background Information
The machine in this activity consists of a single gear train with one driver gear, one driven gear, and two or three idler gears. The relationship (ratio) of teeth on the driver gear to the teeth on the driven gear determines the rate of rotations of the gears. If there are eight teeth on the driver gear and 24 on the driven gear, the driver will turn three times the rate of the driven gear because it has one-third the number of teeth.

The number of gears in a train determine the direction the gears will turn. If there is an even number of gears in the train, the driver and driven gears turn in opposite directions. An odd number of gears allows the driver and driven gears to turn in the same direction.

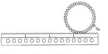

Management

1. This activity works well as an assessment of student understanding of gears. It should be preceded by the other gear activities. Encourage students to make an organized record of their observations of the mystery machine. Also have students record their thinking process and ideas they used as they replicate the machine.

2. This open-ended problem takes a wide range of time for students to solve. Be aware of this time difference and make plans to be flexible.

3. This activity works well as a center with the mystery machine available for examination and several LEGO DACTA® kits available for construction. If this is not possible, the teacher should construct several mystery machines that can be located at different places in the room for students to examine while they try to replicate the machines at their seats.

4. The teacher will need to construct the mystery machines before class. There are six arrangements of driver and driven gears available if gears of the same size are not used, 8:40, 40:8, 8:24, 24:8, 24:40, 40:24. (The 8:40 arrangement is used in the illustration.) The combination, number, and size of idler gears will change depending on what driver and driven gears are chosen.

5. To build the mystery machine, the cover should be copied and cut out. The input and output axles are inserted through the cover's dial faces, the folded cover is wrapped around the machine and taped by the tabs to enclose the model.

6. Rotational motion is typically measured in degrees. The preceding activities measured rotation as a fraction or a decimal of a whole rotation. The extra dial faces are provided for these types of measurement. If it is more appropriate, the teacher should copy these faces and apply them to the machine.

Procedure

1. Show the students the *Mystery Machine* and pose the problem.

2. Encourage the students to examine the machine externally trying to deduce what is inside by the turns of the input and output dials. Suggest that they keep a record of their observations.

3. Ask students to hypothesize about the workings of the machine and try to replicate it with their LEGO® elements. Have them keep a record of their thinking.

4. Direct students to test their machine against the *Mystery Machine* to see if they have been able to replicate it.

5. If the replication does not produce the same results, direct the students to reexamine the *Mystery Machine* and adjust theirs until the replication works.

Discussion

1. How did you determine what gears were inside the *Mystery Machine*?

2. How did you decide how to construct your machine?

3. How does the output dial turn differently than the input dial.

4. What gear arrangements could be used to make the dials do this?

5. How can you tell how many gears are needed in the gear train?

6. Do the number and size of idlers change the input and output dials' turns? Explain. [number affects rotational direction, size of idlers does not affect turns]

7. Does the order of the idlers change the input and output dials' turns? [no] Explain how you know this.

Extensions

1. Have student groups build *Mystery Machines* with other gear arrangements and exchange them to see if others can determine how they are built.

2. Have students build their own original *Mystery Machines*. Have them use any type of linkages they want: gears, beveled gears, chains, belts and pulleys, or U-joints. Have them cover them leaving only the input and output axles exposed, and exchange them to see if others can determine how they are built.

Gear Guessing

Parts Inventory

2 x
5 x
2 x
3 x
2 x
3 x — 4
2 x — 6
2 x
8 x
4 x
4 x
2 x
1 x

1

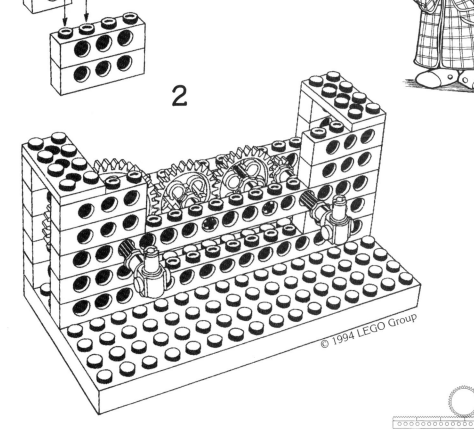

© 1994 LEGO Group

2

© 1994 LEGO Group

Gear Guessing

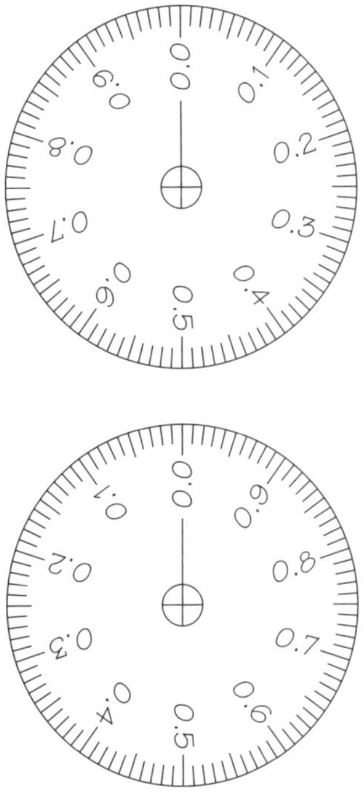

Tab

Tab

The Mystery Machine

Tab

Tab

Decimal Dials

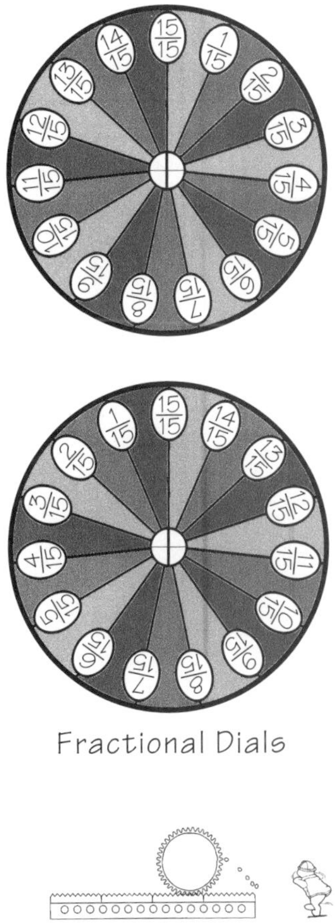

Fractional Dials

Cycloids

A cycloid is a special curve traced by a point on the edge of a circle as the circle is rolled along a straight line. The cycloid was recognized and named by Galileo Galilei. A point located on the rim of a bicycle wheel as it rolls down the street forms a cycloidal arch.

Cycloid

The shape of a cycloid changes as the point on the circle that is being traced is moved into or outside the circle.

- When the point is moved inside the circle, the cycloid is called a prolate cycloid. A prolate cycloid is flatter than a normal cycloid. As the point is moved nearer to the center of the wheel, the arch becomes flatter and flatter.

Prolate Cycloid

- When the point is located on the axle, there is no vertical movement at all.

- When the point is located on an extension outside the circle, the cycloid drawn is called a curate cycloid. A curate cycloid has a loop at the bottom. This loop is below the line along which the circle is being rolled along. The point moves backwards at the bottom of each arch, even though the wheel is moving forward.

Curate Cycloid

If the circle making the cycloid moves forward at a constant speed, the object traveling on the cycloidal path is always changing speeds. At the bottom of the cycloid, the object momentarily stops. The object reaches its highest speed when it reaches the peak of the cycloid, when the circle has made half of a rotation. You may have observed this if you have been in a car at night and a bicycle with reflectors crossed in front of you. The reflectors on the wheels of the bicycle appear to lurch forward as they are constantly changing speeds. Artists indicate this motion when they draw pictures of wheels in motion. The areas below the axle are drawn more distinctly because this area is moving slower or has stopped. A bug that is hitching a ride on the side of the wheel is given a carnival-like ride as it lurches along—moving slowly or stopping at the bottom of the ride and accelerating to high speeds at the top.

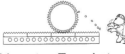

Cycloids are interesting when they are turned upside down and used as tracks or ramps. If two skateboarders are at different heights on either side of a cycloidal ramp, and start down at the same time, they will get to the center of the ramp at the same time. The one farther from the center is on a steeper part of the cycloid. The steeper slope causes this skateboarder to speed up faster than the person at the lower level. The person starting higher reaches a higher speed and is able to cover more distance than the lower, slower starting person.

Skateboarders collide

Finish

Although a straight line is the shortest distance between two points, an inverted cycloid is the fastest path between two points. If a straight ramp and cycloidal ramp run next to each other, and a skateboarder starts down each track at the same time, the next time the tracks meet, the person on the cycloidal ramp will be ahead—even though they started in the same place!

The mathematics involved in the measures of a cycloid are very interesting. The area underneath the arch of a cycloid and above the line is three times the area of the circle used to draw it. Christopher Wren, a seventeenth-century English architect, determined that the length of the cycloid is four times the diameter of the circle used to draw it. He used the cycloid in many of his designs.

The cycloid forms the strongest arch. Many bridges and concrete viaducts are cycloidal arches. Mechanical engineers have applied the cycloidal shape to gears. By having gear teeth cut with cycloidal sides, the teeth are able to make rolling contact as they mesh, thus reducing friction.

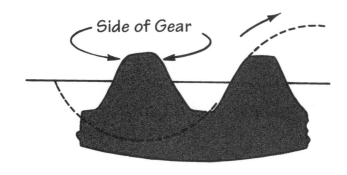

Side of Gear

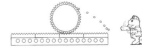

Topic
Geometric shapes–cycloid

Key Questions
1. What would the ride of a bug be like if it were on the rim of your bicycle as you rode down the street?
2. How would its ride be different if it were at a different place on the wheel?

Focus
Students will use a LEGO® gear and racks to draw and learn about the shapes of different types of cycloids. They will make a cycloidal ramp to see an interesting application of the cycloid. This is the first in a series of activities on cycloids and provides an introduction and an interesting application.

Guiding Documents
NCTM Standards
- *Identify, describe, compare, and classify geometric figures*
- *Understand and apply geometric properties and relationships*

Project 2061 Benchmarks
- *Some shapes have special properties: Triangular shapes tend to make structures rigid, and round shapes give the least possible boundary for a given amount of interior area. Shapes can match exactly or have the same shape in different sizes.*

Math
Using geometry
   cycloid

Integrated Processes
Observing
Comparing and contrasting
Generalizing

Materials
LEGO® elements (per group):
   1   1 x 16 beam
   1   gear, 40-tooth
   4   gear racks
Colored pencils
Student pages
Paper strips
Scissors
Cardboard
Straight pins

Masking tape
BBs

Background Information
A *cycloid* is the curved path traced by a point on the circumference of a circle as the circle is rolled along a straight line. The cycloid was first recognized by Mersenne and Galileo Galilei in 1599.

If the path being traced is made from a point on the interior of the circle, the cycloid's form is modified, and this modified shape is called a *prolate cycloid*. If the point is placed on an extension on the exterior of the circle, the cycloid formed is call a *curate cycloid*. A curate cycloid has a loop that is below the line it is rolled along. (The LEGO® beam used as the line in this activity interferes in drawing a curate cycloid.)

Cycloids have some interesting characteristics when they are inverted and used as tracks or ramps. If two BBs which are placed at different distances from the center of an inverted cycloid are released at the same time, they will get to the center at the same moment. The BB farther from the center is on a steeper part of the cycloid. The greater slope causes this BB to accelerate faster than the lower BB. This higher BB reaches a greater speed more quickly and is able to cover more distance than the lower, slower accelerating BB.

Although a line is the shortest distance between two points, an inverted cycloid is the fastest path between two points. When an inverted cycloid track is placed next to a straight track, and a BB is released at the same time on each track from the "start" position, the BB on the inverted cycloid track will be the first to reach the "finish" position.

As the strongest arch, the cycloid has applications in structural engineering. Many bridges and concrete viaducts are cycloidal arches. Mechanical engineers have applied the cycloidal shape to gears. By having gear teeth cut with cycloidal sides, the teeth are able to make rolling contact as they mesh thus reducing friction.

Management
1. This activity is best done in pairs. To draw the cycloid with the pencil in the gear's teeth, one person will need to hold the pencil, while the other person rolls the gear down the rack.
2. This activity has two parts. The first deals with the general shape of the cycloid. The second part has students build a small cycloidal ramp which allows

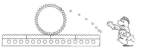

students to observe an interesting aspect of the cycloid. Depending on the students, you may choose to have the students work through only the first part. Younger students may have difficulty controlling the start of the BBs in the second part. In this case, you may want to have the students do the *Extension*.

3. Allow a 45-minute period for each section.
4. Monitor the students as they do their drawings. It is not essential that all cycloids be drawn with the lowest point being at the beginning of the drawing, but it does help students in making comparisons. If necessary, help students position the gear at the start so the lowest point of the cycloid is at the beginning of the their drawings.

Procedure
1. Discuss the *Key Question*.

Part 1 - Shape
2. Distribute the materials necessary for this section: LEGO® elements, student sheet, and colored pencils.
3. Direct the students to attach the four gear racks to the 1x16 beam and place it in position on the student page.
4. Have them put a colored pencil (the "bug") in the center axle slot of the 40-tooth gear and place the gear so the pencil is centered over the left edge of the rack and the gear and rack teeth are meshed.
5. Direct the students to roll the gear down the rack while the pencil records the "bug's trip" in this position.
6. Have the students repeat steps four and five while each time moving the pencil's position farther from the center of the gear. Urge students to use a different colored pencil for each position. Inform students that they need to rotate the gear so that each time the pencil starts as close as possible to the left edge of the rack. When students have completed the record, there should be five curves: one for the center axle slot, one for the off-center axle slots, one for the inner axle hole, one for the outer axle hole, and one for the slots between the gear teeth.

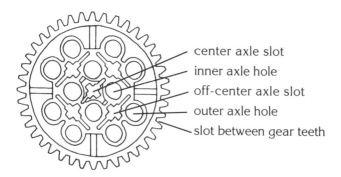

center axle slot
inner axle hole
off-center axle slot
outer axle hole
slot between gear teeth

7. Have students record and discuss observations about the "bug rides" on different positions on the wheel.

8. Explains to students that the geometric shapes they drew are called *cycloids*. If appropriate, introduce the terms *prolate* and *curate*.

Part 2 -Cycloidal Ramp
9. Distribute the second student sheet and have students discuss the question at the top of the sheet.
10. Distribute the necessary materials: LEGO® elements, cardboard, tape, pins, scissors, and BBs.
11. Direct the student to draw a cycloid with the 40-tooth gear at the bottom of the student sheet.
12. Tell them to cut off the bottom of the sheet containing the cycloid.
13. Have students tape the bottom of the student sheet to a piece of cardboard so that both the bottom edge of the cardboard and the bottom of the sheet are matched.
14. Have them put straight pins into the cardboard along the line of the cycloid. The pins should be placed about 0.5 cm apart.
15. When the cycloid is full of pins, have the students cut a strip of paper about one centimeter wide and longer than the cycloid.
16. Have them turn the cycloid upside-down.To form a smooth surface for the cycloidal ramp, direct the students to lay the paper strip on the top of the pins. Have them take a strip of tape and put it on the bottom of the pins. Instruct them to squeeze the tape and paper strip together to hold the paper to the pins. Warn them to be careful not to poke themselves with the pins.
17. Guide one student per group to set the cardboard on a table, holding it so that the cardboard is perpendicular to the table's surface.
18. With the cardboard in this position, direct another student to place a BB on each side of the ramp. Tell the students to release the BBs at the same time and observe what happens, asking them to note where the BBs collide. Direct them to repeat this step several times with the BBs at different heights from each other and from various heights on each trial.
19. Have students discuss their observations. (Students should observe that no matter from what height the BBs were released, they collided in the center or lowest point of the inverted cycloidal ramp.)

Extension
Have students use the cycloids drawn in this activity as models. Have them come up with a method to enlarge the cycloids and build a track for small cars or marbles. They might enlarge it using a grid method or an overhead projector. The cycloid can be cut from cardboard which is used to support the track. Commercial car track materials are available. Cut plastic tubing or hose can be used for marble tracks.

BUG ON A ROLL

What would a bug's ride be like if it were on the rim of your bicycle wheel as you rode down the street?

Use your LEGO® elements to see if you can answer this question by doing the following:

1. Attach four racks to a 1 x 16 beam and place it on the position below.

2. Put a colored pencil in the center axle slot of a 40-tooth gear. Mesh the gear and rack teeth so the pencil is centered over the left edge of the rack.

3. Roll the gear down the rack while the pencil records the "bug's ride."

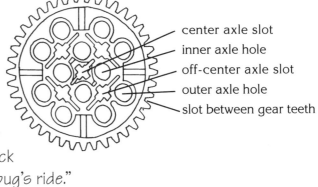

center axle slot
inner axle hole
off-center axle slot
outer axle hole
slot between gear teeth

4. Change to another colored pencil and place it in a position farther from the center of the gear.

5. Rotate the gear so the pencil starts as close as possible to the rack. Mesh the gear and rack teeth and roll the gear so you draw another "bug's ride."

6. Use a different colored pencil to draw a "bug's ride" for each position on the gear. When you are finished, there should be five curves: one for the center axle slot, one for the off-center axle slots, one for the inner axle holes, one for the outer axle holes, and one for the slots between the gear teeth.

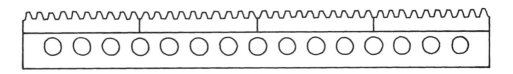

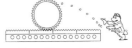

Building a Cycloidal Ramp

What happens when the BBs are released at the same time from different heights and sides of the ramp?

1. Use the 40-tooth gear to draw a cycloid at the bottom of this sheet. Cut off the bottom of the sheet containing the cycloid.

2. Tape this cut-off portion to a piece of cardboard so that both the bottom edge of the cardboard and the bottom edge of the sheet are matched.

3. Put straight pins into the cardboard along the line of the cycloid. The pins should be placed about 0.5 cm apart.

4. Cut a strip of paper about one centimeter wide and a little longer than the cycloid.

5. To make a ramp, turn the cycloid upside-down. Lay the strip of paper on the top of the pins. Secure the paper to the pins by placing a strip of tape on the bottom side of the pins and squeezing the tape and paper strip together.

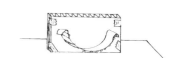

6. Set the cardboard on a table holding it so that the cardboard is perpendicular to the table's surface.

7. While holding the cardboard in this position, have your partner place a BB on each side of the ramp. Release the BBs at the same time and observe what happens. Note where the BBs collide. Repeat this step several times with the BBs at different heights from each other and from various heights on the cycloid.

8. Apply what you observed to what it would be like to ride a bike on a ramp that was shaped like a cycloid.

BRICK LAYERS   115   © 1996 AIMS Education Foundation

BUG - A - LONG

Topic
Geometric Shapes: Length and Area

Key Questions
1. What would the path of a bug's ride be like if it were on the rim of your bicycle wheel as your rode down the street?
2. How is the length of the ride related to the size of the wheel?

Focus
Students will use a LEGO® gear and racks to explore the relationship of the length and area of a cycloid compared to the circle used to draw it. This activity will provide students with experience in developing a concept of area. It develops a formula similar to that of the area of a circle in its relationship to its diameter.

Guiding Documents
NCTM Standards
- *Identify, describe, compare, and classify geometric figures*
- *Understand and apply geometric properties and relationships*
- *Extend their understanding of the concepts of perimeter, area, volume, angle measure, capacity, and weight and mass*

Project 2061 Benchmarks
- *Some shapes have special properties: Triangular shapes tend to make structures rigid, and round shapes give the least possible boundary for a given amount of interior area. Shapes can match exactly or have the same shape in different sizes.*

Math
Using geometry
Measuring
   length
   area
Using proportional reasoning

Science
Physical science
   simple machines
      wheel and axle
   motions

Integrated Processes
Observing
Collecting and recording data
Interpreting data
Comparing and contrasting
Applying and generalizing

Materials
LEGO® elements (per group):
  1 1 x 16 beam
  1 gear, 40-tooth
  4 gear racks
Metric tape measures
Colored pencils
Student pages

Background Information
A cycloid is the curved path traced by a point on the circumference of a circle as the circle is rolled along a straight line.

The area underneath the arch of a cycloid and above the line used to draw it is three times the area of the circle used to draw it. The length of the cycloid is four times the diameter of the circle used to draw it. Christopher Wren, a 17th century English architect, determined this and used the cycloid in constructions.

Management
1. This activity is best done in pairs. To draw the cycloid with the pencil in the gear's teeth, one person will need to hold the pencil, while the other person rolls the gear down the rack.
2. Allow a 45-minute period for this activity.
3. Monitor students as they do their drawings. It is not essential that all cycloids be drawn with the lowest point being at the beginning of the drawing, but it does help students in making comparisons. If necessary, help students position the gear at the start so the lowest point of the cycloid is at the beginning of their drawings.

Procedure
1. Discuss the *Key Question.*
2. Distribute the student page and necessary materials: LEGO® elements, colored pencils, tape measures.
3. Have the students attach the four gear racks to the 1 x 16 beam and place it in position on the student page.
4. Direct them to place the 40-tooth gear on the rack with the gear centered over the left edge of the rack.
5. Guide students to place a colored pencil in the indent between the teeth nearest to the rack and draw a cycloid.
6. Ask them to repeat the procedure using different colors of pencils for both the 24-tooth and eight-tooth gears.

7. Direct them to trace the three gears onto the grid on the student sheet.
8. Have them use a metric tape measure to determine the diameter of each gear and the length of its corresponding cycloid to the nearest millimeter. Have them record the measurements on the chart.
9. Instruct students to count the square units of area of each of the gears and the area underneath each of the corresponding cycloids. Inform them that they will need to do some estimating as to what is considered a full square. Have them record the measurements on the chart.
10. Direct students to calculate the ratios comparing the lengths and areas of the cycloids to the corresponding gears.

Discussion

1. How do the ratios for length compare? [all are close to four]
2. What do the ratios relate about the length of the cycloid compared to the circle that drew it? [The cycloid's length is four times greater than the circle's diameter.]
3. How do the ratios for area compare? [all are close to three]
4. What do the ratios relate about the area under the cycloid compared to the area of the circle that drew it? [The area under the cycloid is three times greater than the circle's area.]

BUG - A - LONG

How far would the bug travel each time the wheel rolled one turn down the street?

Place your beam with the gear racks below and draw a cycloid using each of the three gear sizes.

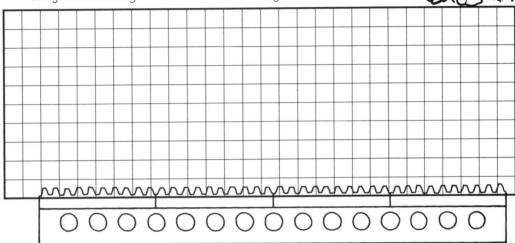

Trace each gear on the grid.

40-tooth	8-tooth	24-tooth

Squares are 0.5 cm on each side. Area of each square equals 0.25 sq. cm.

Make measurements from your drawings to complete the charts

Cycloid Length

Gear Size	Gear Diameter	Cycloid Length	Ratio: $\dfrac{\text{Length}}{\text{Diameter}}$
8-tooth			
24-tooth			
40-tooth			

Cycloid Area

Gear Size	Gear Area	Area Under Cycloid	Ratio: $\dfrac{\text{Cycloid}}{\text{Gear}}$
8-tooth			
24-tooth			
40-tooth			

Speed Bug

Topic
Cycloids; Acceleration

Key Questions
1. What would the ride of a bug be like if it were on the rim of your bicycle as you rode down the street?
2. What would the rate of its turn be like?

Focus
Students will use a LEGO® gear and racks to draw cycloids. With measurements from the cycloids, they will construct graphs to see how the ride on a wheel along a cycloidal path causes a continual change in rate for the object taking the ride. This is a relatively difficult concept and this activity provides a challenge for the able student.

Guiding Documents
NCTM Standards
- *Develop the concept of rates and other derived and indirect measurements*
- *Understand and apply reasoning processes, with special attention to spatial reasoning and reasoning with proportions and graphs*

Project 2061 Benchmarks
- *The graphic display of numbers may help to show patterns such as trends, varying rates of change, gaps, or clusters. Such patterns sometimes can be used to make predictions about the phenomena being graphed.*
- *Things that change in cycles, such as the seasons or body temperature, can be described by their cycle length or frequency, what the highest and lowest values are, and when they occur. Different cycles range from many thousands of years down to less than a billionth of a second.*

Math
Using geometry
   cycloids
Measuring
Graphing
Using rational numbers

Science
Physical science
   rates
   acceleration

Integrated Processes
Observing
Collecting and recording data
Interpreting data
Comparing and contrasting
Applying and generalizing

Materials
LEGO® elements (per group):
  1   1 x 16 beam
  1   gear, 40-tooth
  4   gear racks
Tape measures
Student pages

Background Information
A cycloid is the curved path traced by a point on the circumference of a circle as the circle is rolled along a straight line.

If the circle making the cycloid moves forward at a constant rate, the object traveling on the cycloidal path is always changing the distance it moves, therefore, the speed of the object changes as it moves along the path. (The use of the word *speed* is not used in the data collection of this activity because the drawing of the cycloid is not timed. In order to relate the change in distance over time, it needs to be assumed that the turning rate of the circle is constant.) At the bottom of the cycloid, the object momentarily stops. In a curate cycloid, the object actually goes backward at the bottom of its trip. The object reaches its greatest distance per fractional turn (speed) when it reaches the peak of the cycloid.

Many students will have observed this phenomenon when they have been in a car at night and a bicyclist rode in front of them. Because the distance covered by the reflectors on the bicycle's wheels is constantly changing, the bicycle appears to lurch forward. Artist indicate this motion when they draw pictures of wheels in motion. Areas below the axle are drawn more distinctly as this area is slowing down or stopped.

Management
1. This activity is best done in pairs. To draw the cycloid with the pencil in the gear's teeth, one person will need to hold the pencil, while the other person rolls the gear down the rack.

BRICK LAYERS

© 1996 AIMS Education Foundation

2. Allow a 45-minute period for this activity.
3. Monitor and assist students if necessary as they do their drawings so that all cycloids are drawn with the lowest point being at the beginning of the drawing.

Procedure

1. Discuss the *Key Question*.
2. Distribute the first student page, LEGO® elements, and metric tape measures.
3. Have the students attach the four gear racks to the 1 x 16 beam and place it in position on the student page.
4. Guide them to place the 40-tooth gear on the rack with the gear centered over the left edge of the rack so that the *Fraction Marker* line appears centered in the center axle slot of the gear.
5. Direct the students to place a pencil in the gear slot at the bottom nearest the start line. Have them draw a cycloid as the gear is rolled along the gear rack.

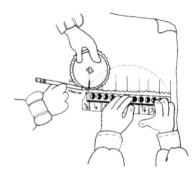

6. When the students have completed the cycloid, tell them to pick up their pencil and gear. Ask them to use the pencil to mark the gear slot they used.
7. Tell them that they are going to repeat the motion of drawing the cycloid, but this time they are going to make a mark on their paper each time a *Fraction Marker* appears in the center axle slot of their gear.

8. When students have finished marking the cycloid, have them use a metric tape measure to find the distance between each mark–the distance traveled by the pencil on the cycloid for each eighth of a rotation the gear made. Direct them to record these measurements under their cycloid drawing.
9. Have them make a bar graph of their measurements to help visualize the changing distance per fractional turn.

10. Distribute the second student page.
11. Have the students transfer the measurements from the first page to the column on the second page *Distance for each 1/8 turn, 1st Rotation*. Tell them to repeat those same measures for *2nd Rotation*.
12. Have them calculate the *Total Distance* (second column) the pencil would travel on the cycloid as the gear made two rotations.
13. Have students make a broken-line graph of the *Total Distance*.

Discussion

1. How are the cycloid and the bar graph similar? [lowest on the ends and high in the middle, symmetrical around center vertical axis]
2. When did the point on the rim go the farthest distance? [near the top of the cycloid, top of the wheel]
3. When is a point on the rim of a wheel going fastest? [near the top of the cycloid, top of the wheel]
4. When did the point on the rim go the shortest distance? [near the ends of the cycloid, bottom of the wheel]
5. When is a point on the rim of a wheel going slowest? [near the ends of the cycloid, bottom of the wheel]
6. How does the change in the rate of turn show up on the coordinate graph? [changing slope or steepness]
7. A cartoon figure has appeared at the bottom of the pages of all the lessons you've done using the LEGO® elements. What relationship can you find between that figure and the last three lessons, *Bug on a Roll, Bug-A-Long,* and *Speed Bug*? [This is a flip-book figure. The headlamp of the cartoon figure forms a cycloid as it turns. The circles of the headlamp leave a record of the distance covered for each sixteenth of a turn.]

Extensions

1. Make a flip book of the cartoon figure and relate it to the lessons on cycloids. (See *Make Your Own Flip Book*.)
2. To bring closure to this series of three activities, have the students write a description of the ride of the bug on the wheel rim of a bicycle. Have them write it as if they are the bug, describing the sensations and observations they see as the wheel rolls down the road. This activity provides a good assessment of student understanding. Before having students do the writing, have them discuss what the ride would be like. This will provide an opportunity to review all the aspects of the cycloid that were studied in the series.
3. Have students gather drawings of moving wheels and identify evidence of the cycloidal motion of the wheel's rim (see *Background Information*).

How does the distance the bug travels on the path change as it rolls down the street?

1. Put the 40-tooth gear on the gear rack with the gear centered over the left edge of the rack so that the *Fraction Marker* line appears centered in the center axle slot of the gear.
2. Place your pencil in the gear slot at the bottom nearest the start line.
3. Draw a cycloid as you roll the gear along the gear rack.
4. Pick up your pencil and gear. Use the pencil to mark the gear slot you used when you drew the cycloid.
5. Repeat the motion of drawing the cycloid, but this time make a mark through the gear slot on the cycloid path each time a *Fraction Marker* appears in the center axle slot of your gear.

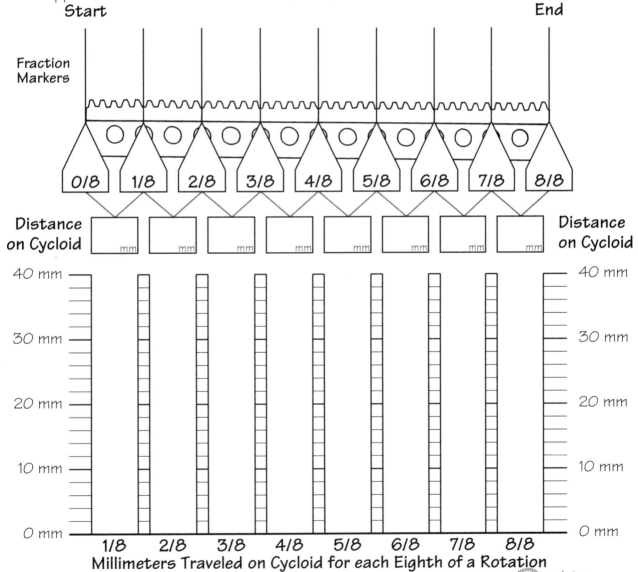

© 1996 AIMS Education Foundation

SPeeD BuG

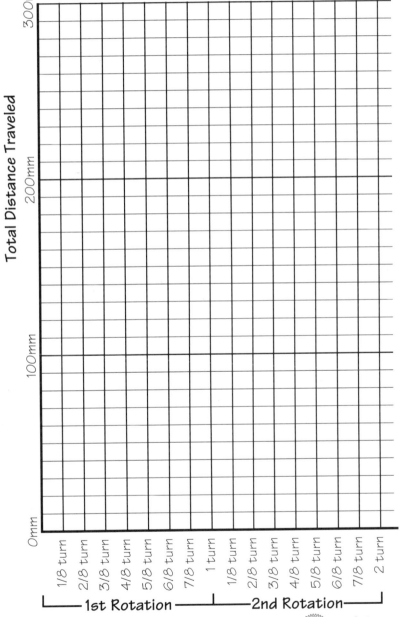

Copy the *Distance on Cycloid* measures from the last page onto the *Distance for each 1/8 turn, 1st Rotation*. Repeat those same measures for *2nd Rotation*. Calculate *Total Distance* traveled. Use the *Total Distance* data to make a broken-line graph. Compare this graph to the bar graph.

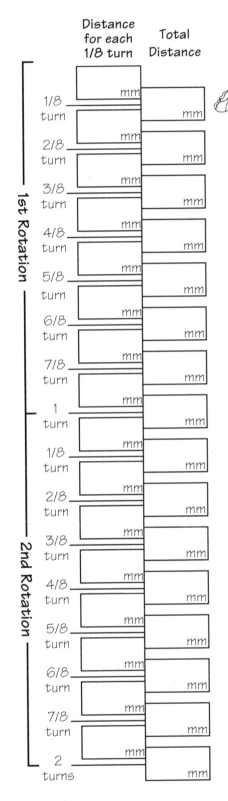

Total Distance Traveled

300mm

200mm

100mm

0mm

1/8 turn · 2/8 turn · 3/8 turn · 4/8 turn · 5/8 turn · 6/8 turn · 7/8 turn · 1 turn · 1/8 turn · 2/8 turn · 3/8 turn · 4/8 turn · 5/8 turn · 6/8 turn · 7/8 turn · 2 turn

├── 1st Rotation ──┤├── 2nd Rotation ──┤

Make Your Own Flip Book

1. Copy onto cardstock or heavy-weight paper.

2. Cut out the 36 rectangles.

3. Stack them in sequential order with number one on the top.

4. Fasten together with one staple along the left edge.

5. Use the right edge to flip through the book.

1	5
2	6
3	7
4	8

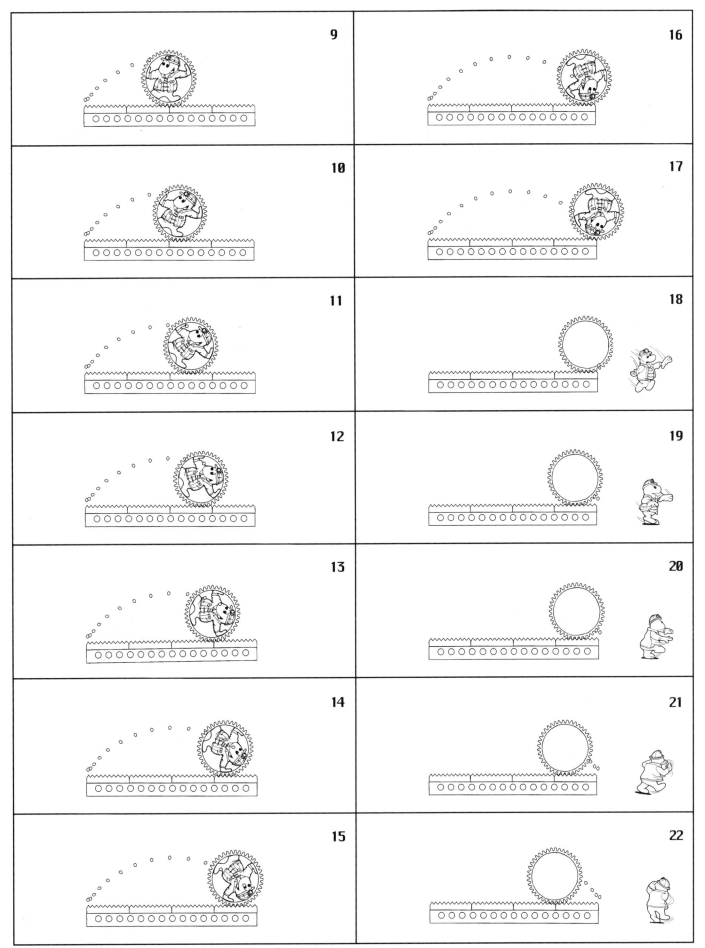

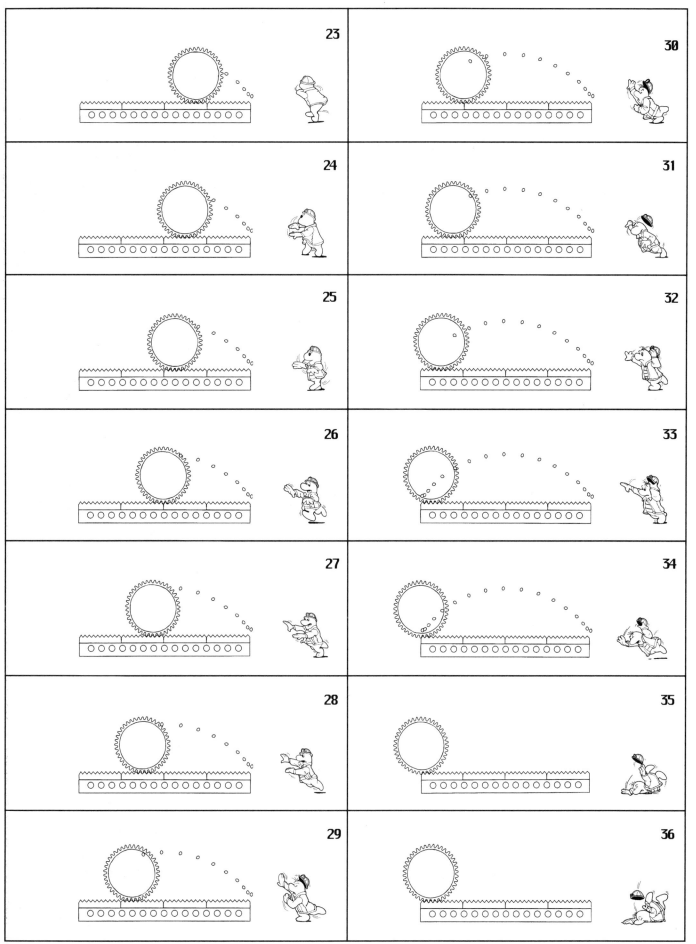

Stable Structures

Forces

All structures are affected by two forces, tension and compression. Understanding how tension and compression work makes it possible to determine how structures stand up.

Tension is a pulling or stretching force. When you pull on a string, you are tensing it or putting it in tension. Tension tends to make things longer– it extends them.

Compression is a pushing force. A vertical wall in a building is under compression. Because of the force of gravity, the weight of the wall is pushing it down. Compression tends to make things shorter.

You can feel both forces if you get together with a partner and stand facing each other with your feet touching the other person's feet. Lock your hands and with your legs straight, both of you lean back. You should feel the tension in your arms as you pull to keep the other person from falling away. You should also feel the compressive force of your weight pushing down on your legs.

Materials

Different materials can deal with different types of forces better than others. Steel, glass, and plastic cables can support a great deal of hanging weight. They have tensile strength–they can be stretched. Since they are flexible, they have relatively small compressive strength. Steel beams, rock, and concrete all have relatively strong compressive strengths. Materials with relatively strong compressive strength are often relatively heavy.

Reinforced concrete combines the strengths of two different materials. It gets compressive strength from the concrete and combines it with the strong tensile strength of steel rods.

Buildings such as houses have both compression and tension acting on them. Due to the force of gravity, the weight of the rafters on the pitched roof pushes them down. They are in compression. Some of this compression is transmitted to the walls and finally pushes down on the Earth. The rest of the force pushes outward on the walls. This force is called thrust. If left this way, this compression would push the walls out and the house would collapse.

You can feel these forces if you work with your partner again. Face each other, raise your hands higher than your heads, and put your palms together with your partner's. Walk backwards slowly, keeping your hands together. Lean on each other until you feel you would fall if your partner let you go. You should feel the compression in your arms and legs. If you continue to walk backwards, you should begin to feel your feet slide out from under you. This is the outward push of thrust.

Making Structures Stable

One way to counter the outward thrust in a building is to place a tie rod between the two ends of the rafters. It would be like tying a rope between you and your partner's ankles.

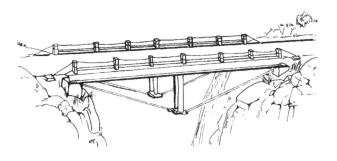

This tie rod is in tension because it pulls and keeps the ends of the rafters from stretching apart. In homes, these tie rods are made of wood and serve as the joists for the ceiling. In other structures, they may be made from steel rods or cables.

Trusses are structures that have some members (parts) in tension, while others are in compression. The rafters and joists of a pitched roof form a truss. To make the most efficient truss, the materials must be matched to the forces that will be applied to them. Parts of a truss that will be in tension should use materials of high tensile strength. Replacing tensed steel beams in a truss with rods or cables reduces the weight and cost of the structure without reducing the strength.

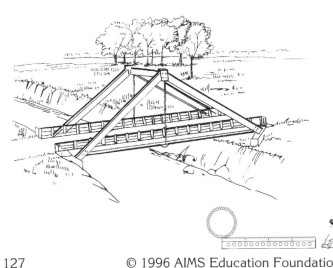

Shapes

Some shapes are more stable than others. The triangle is the only polygon that is intrinsically stable (does not change shape). If one vertex is moved, the other two vertices must move with it. The length of one side cannot be changed without changing at least one of the others. This unique stability makes the triangle very useful in construction. It is used to form stable structures and to reinforce unstable shapes.

Other polygons can be modified by connecting the vertices with braces. The braces divide the polygon into triangles, the more stable shape. The rectangular walls of buildings are flimsy until cross-braces are nailed at a diagonal. The diagonal cross-bracing produces two stable triangles and keeps the wall from changing shape.

In many constructions such as bridges, these stabilizing braces can be seen. In other buildings such as skyscrapers, it is hidden under a superficial skin. In houses, where forces that try to change the shape of the wall are relatively small, the material used as the surface of the wall may be the brace. When unusual forces such as earthquakes, act on structures which have not been stabilized well, they collapse much more easily.

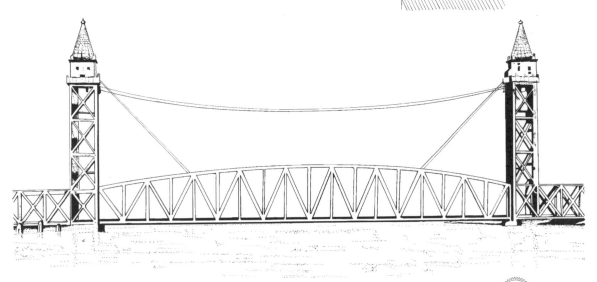

A STABLE TABLE

Topic
Structures; Polygons

Key Question
What polygons can be used to make a stable structure?

Focus
Students will build polygons and test them to see which ones are stable.

Guiding Documents
NCTM Standards
- *Apply mathematical thinking and modeling to solve problems that arise such as art, music, psychology, science, and business*
- *Identify, describe, compare, and classify geometric figures*

Project 2061 Benchmarks
- *Some shapes have special properties: Triangular shapes tend to make structures rigid, and round shapes give the least possible boundary for a given amount of interior area. Shapes can match exactly or have the same shape in different sizes.*
- *Design usually requires taking constraints into account. Some constraints, such as gravity or the properties of the materials to be used, are unavoidable. Other constraints, including economic, political, social, ethical, and aesthetic ones, limit choices.*

Math
Using geometry
   polygons

Science
Physical science
   forces
   structures
      stability

Integrated Processes
Observing
Comparing and contrasting
Collecting and recording data
Generalizing
Applying

Materials
LEGO® elements (per group):
- 4   1 x 16 beams
- 2   1 x 12 beams
- 2   1 x 8 beams
- 8   connector pegs

Background Information
The names of the polygons in this activty are:

Sides	Name
3	Triangle
4	Quadrilateral
5	Pentagon
6	Hexagon
7	Heptagon
8	Octagon

The triangle is the only polygon that is intrinsically stable (does not change shape). If one vertex is moved, all its other vertices must move with it. The length of one side cannot be changed without changing at least one of the others. This unique stability makes the triangle very useful in construction. It is used to form stable structures and to reinforce unstable shapes.

Management
1. This activity can be completed in 20 to 30 minutes.
2. It is best done in groups of two.
3. Before doing the activity, it may be necessary to demonstrate how the beams can be connected with a connector peg to make a movable joint.
4. Students will need to know the names of polygons.

Procedure
1. Discuss the *Key Question*.
2. Ask students to take three beams and connectors pegs to form a triangular shape.
3. Have them place the structure so it lies flat on the table. Tell them to hold one of the beams securely to the table while they push and pull on the shape to determine if it is stable (does not change shape).
4. Direct the students to record the number of sides, the name of polygon, and if it is stable or unstable.

5. Have students repeat this procedure after adding another beam to the polygon. Have them continue adding beams until they have completed the octagon. (The polygons formed will not be regular since beams of different lengths are used.)

Discussion
1. Which polygons are stable? [triangles]
2. Which polygons are not stable? [all but the triangle]
3. If you were going to make a structure and did not want it to be unstable and move, what polygons would you use in your construction? [triangles]
4. Explain in your own words why you think triangles are stable.
5. Where have you seen triangles used to make structures more stable?

Extensions
1. Have students make a list of different structures they observe around them. Have the students identify the geometric shapes and make inferences as to why the structure is made with this shape.
2. Have students explore ways to make unstable polygons stable (see *Angle Fixer*).

A STABLE TABLE

Which polygons are stable?

Polygon						
Number of Sides						
Name of Polygon						
Stable or Unstable						

What conclusion can you make from the information in this table?

BRICK LAYERS

131

Topic
Structures; Polygons

Key Question
How could you modify a polygon to make it stable?

Focus
Students will modify polygons to make them stable.

Guiding Documents
NCTM Standards
- *Describe and represent relationships with tables, graphs, and rules*
- *Identify, describe, compare, and classify geometric figures*

Project 2061 Benchmarks
- *Some shapes have special properties: Triangular shapes tend to make structures rigid, and round shapes give the least possible boundary for a given amount of interior area. Shapes can match exactly or have the same shape in different sizes.*
- *Engineers, architects, and others who engage in design and technology use scientific knowledge to solve practical problems. But they usually have to take human values and limitations into account as well.*

Math
Using geometry
    polygons
Identify patterns
Using and applying formula

Science
Physical science
    structures
        stability
    forces

Integrated Processes
Observing
Comparing and contrasting
Collecting and recording data
Generalizing
Interpreting data

Materials
LEGO® elements (per group):
  4   1 x 16 beams
  2   1 x 12 beams
  2   axles, 12-studs long
  1   axle, 10-studs long
 12   connector pegs
  6   piston rods

Background Information
The triangle is the only polygon that is intrinsically stable (see *A Stable Table*). However, all polygons can be modified by connecting the vertices which divide the interior of the polygon into triangular regions. The least number of connecting members needed to stabilize the polygon is always three less than the number of sides of the polygon. The number of regions the polygon is divided into is two less than the number of sides of the polygon.

The quadrilateral is one of the most common polygons used in construction. It is considered an unstable shape because a force applied to any of the four sides will cause the shape to change. To stabilize this shape, one diagonal connection is made between two opposite corners. This reinforcing member divides the quadrilateral into two triangles.

In many constructions (bridges, for example), this stabilizing member can be seen. In other constructions such as skyscrapers, it is hidden under a superficial skin. In constructions, such as those in a house where forces that deform the quadrilateral are relatively weak, the surface material of the wall may be the only stabilizing member. When unusual forces such as earthquakes act on constructions which have not been stabilized well, they collapse much more easily.

Management
1. Allow 45 minutes to complete this activity. This time may have to be increased if done as an open-ended exploration.
2. This activity is done best in groups of two.
3. If necessary, demonstrate for the students how axles can be used as stabilizing members when connected to the beams with piston rods and connectors.

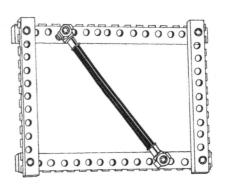

4. This activity works well as an open-ended exploratory investigation. If you choose this option, discuss the *Key Question* with the students and provide time for them to explore, gather information, and draw conclusions.

5. Not all regions in the stabilized shape will be triangles, but students should recognize the need for triangles for some regions.

Procedure

1. Discuss the *Key Question*. Review the findings from *A Stable Table*.

2. Direct students to build the quadrilateral out of beams and connector pegs.

3. Have them use an axle with piston rods and connector pegs to make a stabilizing member which diagonally connects two corners.

4. Instruct the students to place the structure so it lies flat on the table. Have them hold one of the beams securely to the table as they push and pull on the shape to determine if it is stable. If it is not stable, direct them to continue to add axles until it is stable. Encourage them to use the fewest number of axles possible to make the shape stable.

5. When students have modified their quadrilateral and have made it stable, have them record a sketch of it on the record sheet along with the number of axles used, and the number of regions into which the original polygon has been divided.

6. Direct them to complete steps 2 to 5 for the pentagon and hexagon.

7. Discuss their generalizations.

Discussion

1. What pattern do you see between the number of sides in the original polygon, and the number of reinforcing axles? [3 less]

2. Write an equation that states this pattern. [Sides - 3 = Reinforcing Axles]

3. What pattern do you see between the number of sides in the original polygon, and the number of regions? [2 less]

4. Write an equation that states this pattern. [Sides - 2 = Regions]

5. What is the general shape of the regions of the stabilized polygons? [triangular]

6. How could you use what you have learned in this exploration to help you build stable structures? [Answers will vary, but students should recognize that if a structure is unstable, it can be stabilized by modifying the structure to contain triangular regions.]

Extensions

1. Have students gather pictures of different structures that have been modified by triangles to make them stable. Have them identify the original, or exterior shape, and the stabilizing members that make triangles.

2. Have students use the shape of one of the stable models as a basis of a structure (bridge, boom, building) and have them make a sketch of the structure.

Angle Fixer

How can you modify a polygon to make it stable?

Original Polygon	Number of Sides	Name of Polygon	Diagram of Stabilized Polygon	Number of Axles Used	Number of Regions

What patterns do you see between the number of sides on the original polygon and the number of axles or number of regions used to stabilize it?

BRICK LAYERS

134

STRESS
on a String

Topic
Structures; Forces–compression, tension

Key Question
How can you build a stable bridge from pieces that are connected together in a flexible way?

Focus
Students will build various trusses and determine which members (parts) are under compression or tension.

Guiding Documents
NCTM Standards
- *Develop and apply a variety of strategies to solve problems, with emphasis on multistep and nonroutine problems*
- *Generalize solutions and strategies to new problem situations*
- *Understand and apply geometric properties and relationships*

Project 2061 Benchmarks
- *Engineers, architects, and others who engage in design and technology use scientific knowledge to solve practical problems. But they usually have to take human values and limitations into account as well.*
- *Design usually requires taking constraints into account. Some constraints, such as gravity or the properties of the materials to be used, are unavoidable. Other constraints, including economic, political, social, ethical, and aesthetic ones, limit choices.*

Math
Geometry

Science
Physical science
   structures
   forces
      tension
      compression

Integrated Processes
Observing
Comparing and contrasting
Applying and generalizing

Materials
LEGO® elements (per group):

4	2 x 4 bricks
2	1 x 4 beams
4	1 x 16 beams
4	piston rods
2	axle, 12-studs long
7	connector pegs

String

Background Information
All structures are affected by two forces, tension and compression. Understanding how tension and compression work makes it possible to determine how structures stand up.

Tension is a pulling or stretching force. When you pull on a string, you are tensing it or putting it in tension. Tension tends to lengthen the member (extension).

Compression is a pushing force. A vertical wall in a building is under compression. Due to the force of gravity, the wall's weight pushes it down. Compression tends to shorten the member.

Different materials can withstand different types of forces better than others. Steel, glass, and plastic cables can support a great deal of hanging weight. They have tensile strength. However, since they are flexible, they have relatively less compressive strength. Steel beams, rock, and concrete all have relatively greater compressive strengths. Materials with greater compressive strength are often relatively heavy. Reinforced concrete is a combination of two different materials. The concrete provides great compressive strength while the internal reinforcing steel rods provide great tensile strength.

The rafters on the pitched roof of a house are affected by gravity. Their weight pushes them down. They are in compression. Some of this compression is transmitted to the walls and finally pushes down on the Earth. The rest of the force pushes outward on the walls. If left this way, the force of compression would push the walls out and the house would collapse. One way to counter this outward thrust is to place a tie rod between the two ends of the rafters. The tie rod is in tension because it pulls and keeps the ends of the rafters from stretching apart with the outward thrust. In homes these tie rods are made of wood and serve as the joists for the ceiling. In other structures, they may be made from steel rods or cables.

Trusses are structures that have some members (parts) in tension, while others are in compression. The rafters and joists of a house with a pitched roof form a truss. To make the most efficient truss, the materials must be matched to the force that will be applied to them. Members of a truss that will be in tension should use materials of high tensile strength. Replacing tensed steel beams in a truss with rods or cables reduces the weight and cost of the structure without reducing the strength.

In trusses made from LEGO® elements, tensed beams may be replaced with string. Using string allows students to take advantage of a limited number of pieces while maintaining a stable structure. Below are two trusses that will be built by students at the conclusion of this activity. The compressed and tensed members are labeled. The trusses will be stable when string is used for the tensed members.

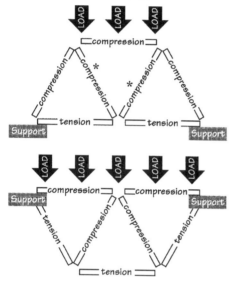

* Under the symmetrical load illustrated, these members are actually zero force members. Since students will not be able to apply a symmetrical load, they will most likely identify them as "under compression."

Management
1. This activity works best in small groups of two.
2. Before doing this activity, instruct students how to make a overhand knot with a loop. An overhand knot at both ends of a piece of string makes a tie rod for the tensed pieces. This knot can be slipped on and off the end of a connector peg. For younger students, it may be necessary to pre-tie the string.
3. Before doing this with the class, build several trusses in order to anticipate any difficulties students may face.
4. If students feel it is easier to solve the final problem by using only beams and axles, a competition can be held to see who can use the least number of beams and axles.

Procedure
1. Discuss the meaning of tension and compression.
2. Pose the *Key Question*.
3. Direct students to build the first two trusses that are illustrated.

4. Have them follow the instructions on the student sheets, make observations of the trusses, and answer the questions.
5. Discuss the observations and generalizations made by the students.
6. Have students predict and record which members of the two trusses diagrammed on the final student page are in compression or tension.
7. Direct students to construct the trusses using beams and axles for compressed members and string for tensed members.
8. Have students validate their predictions or correct the truss until it is stable.

Discussion
1. How can you tell that a string is in tension? [It is taut.]
2. What shape is consistently found in stable trusses? [triangle]
3. What happened to each of the stable trusses as you turned them upside down? [They folded up.]
4. Why doesn't a truss remain stable in the inverted position? [When inverted, the members that were in tension turn into members in compression. String has no compressive strength.]
5. What generalization can you make about the position of members in tension? [Tensile members are found on the bottom of trusses where the outward thrust of diagonal compressive members must be countered.]
6. What are some applications for flexible trusses? [buildings, bridges, booms, or other structures where forces are applied consistently in one direction]
7. Why is a large radio antenna held by three cables, each going in a different direction? [Wind, the force acting against an antenna, can blow from any direction. The cables provide the tension countering the wind.]

Extensions
1. Have students watch for and make a sketch of different trusses they see. Have them determine which members are in tension and which are in compression.
2. Have students use LEGO® elements to make the frame of the house shown below. Challenge them to make the frame support the weighted brick by only adding string and connector pegs.

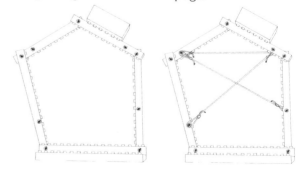

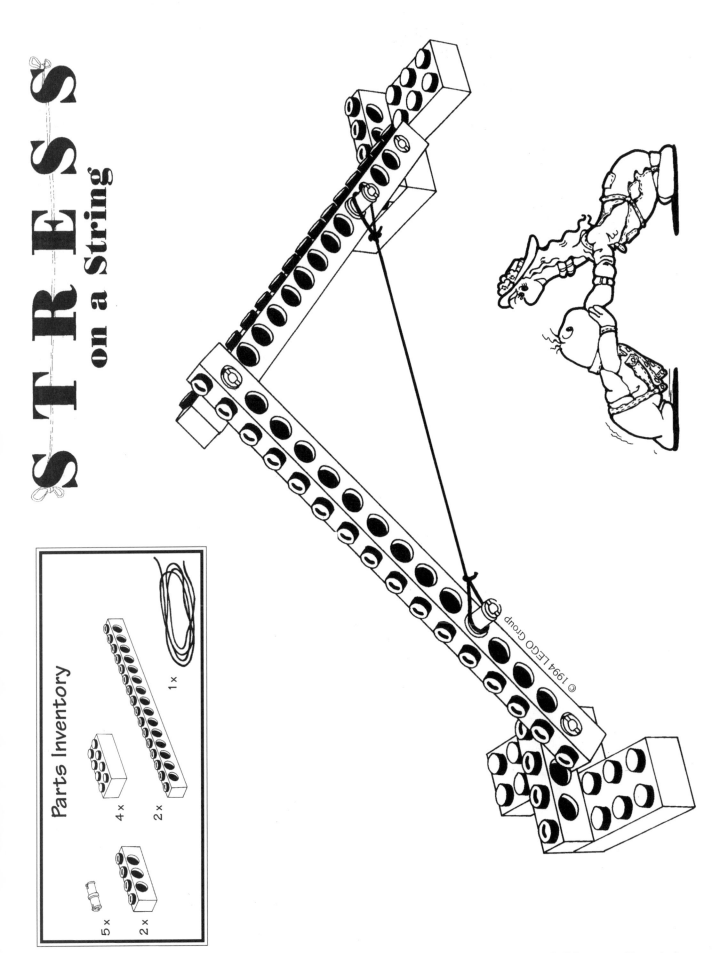

STRESS
on a String

Parts Inventory

5 x
2 x
4 x
2 x
1 x

© 1994 LEGO Group

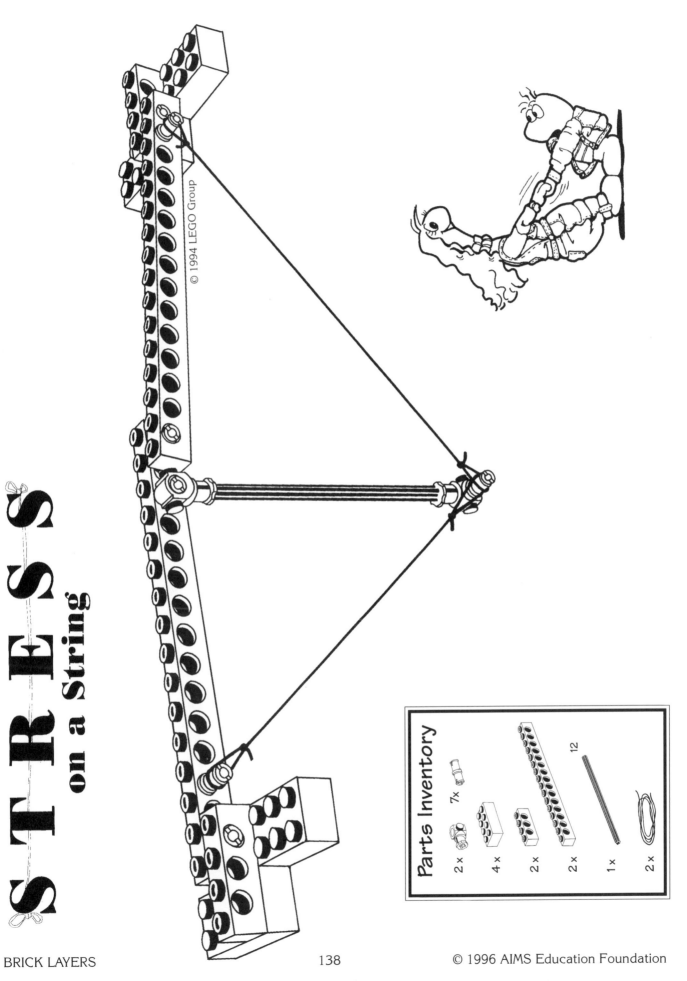

S T R E S S
on a String

Parts Inventory

Part	Quantity
	7 x
	2 x
	4 x
	2 x
	2 x
12	1 x
	2 x

© 1994 LEGO Group

S T R E S S
on a String

Use LEGO® elements and string to build the truss in the illustration below.

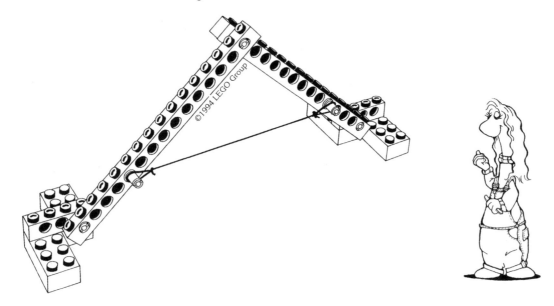

Put two desks near each other and place the truss so it spans the distance between the edges of the desks. Push down on the joint of the two beams.

1. Was the truss stable before you set the supports on the table?

2. Is the truss stable as you push down on the joint?

3. Feel the string as you push down on the joint. What happens to the string?

4. What force is evident on the string?

5. Label the illustration to show the members that are in compression and the members that are in tension.

6. Turn the truss upside down from what is illustrated. Explain what happens to the truss and why it happens.

S T R E S S
on a String

Now build the truss illustrated below.

©1994 LEGO Group

Place the supports of this truss on the two desks so that the strings and strut are suspended underneath the two beams. Push down on the joint of the two beams.

7. What happens to the strings as you push on the joint?

8. Label the illustration to show the members that are in compression and the members that are in tension.

9. Turn the truss upside down from what is illustrated so that the string and strut are above the two beams. Explain what happens to the truss and why it happens.

STRESS
on a String

In LEGO® constructions string can be used for a member in tension. Actual structures often use steel rods or cables to replace the tensed members. Using the rods or cables reduces the weight of the structure and the cost, without reducing the strength. Using string in LEGO® constructions allows you to take advantage of a limited number of pieces while maintaining a stable structure.

As you have observed from the two trusses you built, it is very important to determine what members are in tension. String can replace only those members that are in tension. If string is used as a member in compression, the structure will collapse.

Below are the diagrams for two trusses. They are identical except in one the truss is placed above the level of support, and in the second it is suspended below the level of support. Before constructing each truss, predict which members are in tension and which are in compression. Use your prediction to label the members in the illustration. Then build the truss using string for all tensile members.

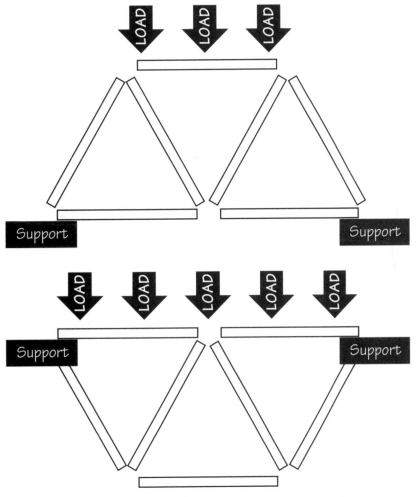

THE BIG BOOM
Construction Project

Topic
Structures and forces

Key Question
Using the contents of one LEGO DACTA® kit and string, how can you make the longest boom that bends the least under the load of a weighted brick?

Focus
Students will apply their understanding of tension, compression, and stability to construct the longest stable boom.

Guiding Documents
NCTM Standards
- *Develop and apply a variety of strategies to solve problems, with emphasis on multi-step and non-routine problems*
- *Discuss mathematical ideas and make conjectures and convincing arguments*
- *Apply mathematical thinking and modeling to solve problems that arise such as art, music, psychology, science, and business*
- *Estimate, make, and use measurements to describe and compare phenomena*

Project 2061 Benchmarks
- *Mathematics is helpful in almost every kind of human endeavor – from laying bricks to prescribing medicine or drawing a face. In particular, mathematics has contributed to progress in science and technology for thousands of years and still continues to do so.*
- *Design usually requires taking constraints into account. Some constraints, such as gravity or the properties of the materials to be used, are unavoidable. Other constraints, including economic, political, social, ethical, and aesthetic ones, limit choices.*
- *The choice of materials for a job depends on their properties and how they interact with other materials. Similarly, the usefulness of some manufactured parts of an object depends on how well they fit together with the other parts.*

Math
Using geometry
Measuring
    linear
Using computation

Science
Physical science
    forces
        compression
        tension
    stability
    deflection

Integrated Processes
Observing
Comparing and contrasting
Applying and generalizing
Evaluating

Materials
LEGO® elements (per group):
    beams
    axles
    piston rods
    connector pegs
    building plate
String
Meter stick or tape measure

Background Information
A *boom* is a beam projecting from the mast of a derrick. It is usually counterbalanced to keep the derrick from tipping over. In this activity, the boom will be held against the top of a table and extended out over the edge.

Deflection is the amount a structure bends under a load. In this activity the deflection will be the difference between the height of the end of the boom without the weighted brick and the height of the end of the boom with the weight.

Refer to *Background Information* in the preceding three activities for information about stability, compression, and tension.

Management
1. Groups of two to four students work well for this activity.
2. Any mass of 50-100 grams may be used if a weighted brick is not available.
3. This is an opened-ended problem so the time required varies greatly. Provide at least one hour for students to construct and modify their booms.
4. Students will often engineer a boom to counter vertical bending. Rarely do they consider the need

for guys to prevent any lateral twisting. Resulting booms may fail by twisting sideways. Be aware of this common weakness and take advantage of it by guiding students to recognize how and why their booms failed. Encourage them to modify the design to overcome the weakness.

5. Often it is best to compress a beam between the building plate and the table to secure the boom. This is "legal" and keeps the boom from lifting off the plate as the load is applied. (See illustration below.)

Procedure

1. Distribute the rules for the contest.
2. Discuss all the rules and how scoring will be done. Make sure that all students understand the information.
3. Provide students time to construct and modify their booms.
4. Have them sketch their boom and write an explanation to the reasoning behind their design.
5. Hold the contest. Have one student place the boom at the edge of the table and secure it by only holding the building plate.
6. Verify that the boom complies with all rules.
7. Direct the students to measure and record the length of the boom from the end of the table to the end of the boom.
8. Inform them that the unloaded height of the end of the boom is measured from the floor and recorded.
9. Have the students attach the weighted brick or suspend it from the end of the boom. Direct them to measure and record the loaded height of the end of the boom from the floor.

10. Have students calculate and record the deflection by subtracting the loaded height from the unloaded height.
11. Have them calculate and record the score by subtracting the deflection from the boom's length.
12. Repeat steps 5-11 for each group.
13. Have students rank the booms by scores.
14. Discuss the differences of design in high and low scoring booms.

Discussion

1. What different methods did your group apply to make sure your boom was stable under a load?
2. Identify members of your boom that are under tension...that are under compression.
3. What is the longest boom?
4. How were the longer booms constructed?
5. Which boom deflected the least?
6. How were beams with low deflections constructed?
7. Which boom got the highest score? Why?
8. How were booms constructed that had the highest scores?
9. If there were to be a rematch, how would you modify your boom?

Extensions

1. Allow students to modify their booms and have a rematch.
2. Have a similar contest for the longest flexible bridge. Suggested rules would include:
   Bridge must be self-supporting.
   No contact below the level of the table.
   No more defection than 5% of length.

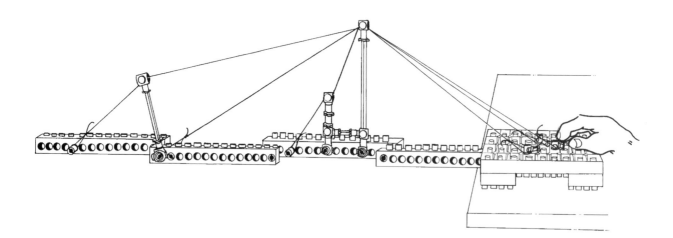

The Big Boom Construction Contest

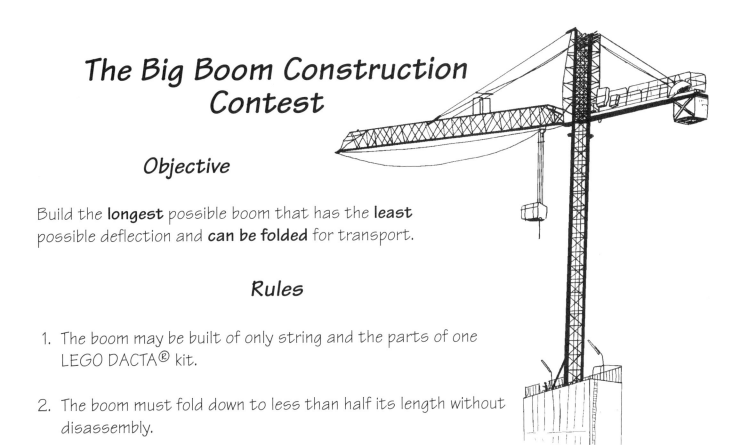

Objective

Build the **longest** possible boom that has the **least** possible deflection and **can be folded** for transport.

Rules

1. The boom may be built of only string and the parts of one LEGO DACTA® kit.

2. The boom must fold down to less than half its length without disassembly.

3. One end of the boom must be attached to a LEGO® building plate as a foundation.

4. One student will secure the boom to the table by holding **only** the building plate.

5. When secured at the edge of the table for judging, no part of the boom may fall below the level of the table.

Judging

1. When the boom has been secured to the table, the boom's **length** (the distance from the table edge to the boom's suspended end) is measured.

2. The **height** of the boom's suspended end is measured from the floor.

3. The boom is loaded with a weighted brick and the **deflected height** of the suspended end is measured.

4. A **score** is calculated by subtracting the deflection (unloaded height minus loaded height) from the boom's length.

5. The boom with the highest score wins the contest.

THE BIG BOOM

Construction Project

Group Members	Boom Length	Unloaded Height -	Loaded Height =	Deflection	Score (Length - Deflection)	Class Rank

The AIMS Program

AIMS is the acronym for "**A**ctivities **I**ntegrating **M**athematics and **S**cience." Such integration enriches learning and makes it meaningful and holistic. AIMS began as a project of Fresno Pacific University to integrate the study of mathematics and science in grades K-9, but has since expanded to include language arts, social studies, and other disciplines.

AIMS is a continuing program of the non-profit AIMS Education Foundation. It had its inception in a National Science Foundation funded program whose purpose was to explore the effectiveness of integrating mathematics and science. The project directors in cooperation with 80 elementary classroom teachers devoted two years to a thorough field-testing of the results and implications of integration.

The approach met with such positive results that the decision was made to launch a program to create instructional materials incorporating this concept. Despite the fact that thoughtful educators have long recommended an integrative approach, very little appropriate material was available in 1981 when the project began. A series of writing projects have ensued and today the AIMS Education Foundation is committed to continue the creation of new integrated activities on a permanent basis.

The AIMS program is funded through the sale of this developing series of books and proceeds from the Foundation's endowment. All net income from program and products flows into a trust fund administered by the AIMS Education Foundation. Use of these funds is restricted to support of research, development, and publication of new materials. Writers donate all their rights to the Foundation to support its on-going program. No royalties are paid to the writers.

The rationale for integration lies in the fact that science, mathematics, language arts, social studies, etc., are integrally interwoven in the real world from which it follows that they should be similarly treated in the classroom where we are preparing students to live in that world. Teachers who use the AIMS program give enthusiastic endorsement to the effectiveness of this approach.

Science encompasses the art of questioning, investigating, hypothesizing, discovering, and communicating. Mathematics is a language that provides clarity, objectivity, and understanding. The language arts provide us powerful tools of communication. Many of the major contemporary societal issues stem from advancements in science and must be studied in the context of the social sciences. Therefore, it is timely that all of us take seriously a more holistic mode of educating our students. This goal motivates all who are associated with the AIMS Program. We invite you to join us in this effort.

Meaningful integration of knowledge is a major recommendation coming from the nation's professional science and mathematics associations. The American Association for the Advancement of Science in *Science for All Americans* strongly recommends the integration of mathematics, science, and technology. The National Council of Teachers of Mathematics places strong emphasis on applications of mathematics such as are found in science investigations. AIMS is fully aligned with these recommendations.

Extensive field testing of AIMS investigations confirms these beneficial results.

1. Mathematics becomes more meaningful, hence more useful, when it is applied to situations that interest students.
2. The extent to which science is studied and understood is increased, with a significant economy of time, when mathematics and science are integrated.
3. There is improved quality of learning and retention, supporting the thesis that learning which is meaningful and relevant is more effective.
4. Motivation and involvement are increased dramatically as students investigate real-world situations and participate actively in the process.

We invite you to become part of this classroom teacher movement by using an integrated approach to learning and sharing any suggestions you may have. The AIMS Program welcomes you!

AIMS Education Foundation Programs

A Day with AIMS®

Intensive one-day workshops are offered to introduce educators to the philosophy and rationale of AIMS. Participants will discuss the methodology of AIMS and the strategies by which AIMS principles may be incorporated into curriculum. Each participant will take part in a variety of hands-on AIMS investigations to gain an understanding of such aspects as the scientific/mathematical content, classroom management, and connections with other curricular areas. *A Day with AIMS®* workshops may be offered anywhere in the United States. Necessary supplies and take-home materials are usually included in the enrollment fee.

A Week with AIMS®

Throughout the nation, AIMS offers many one-week workshops each year, usually in the summer. Each workshop lasts five days and includes at least 30 hours of AIMS hands-on instruction. Participants are grouped according to the grade level(s) in which they are interested. Instructors are members of the AIMS Instructional Leadership Network. Supplies for the activities and a generous supply of take-home materials are included in the enrollment fee. Sites are selected on the basis of applications submitted by educational organizations. If chosen to host a workshop, the host agency agrees to provide specified facilities and cooperate in the promotion of the workshop. The AIMS Education Foundation supplies workshop materials as well as the travel, housing, and meals for instructors.

AIMS One-Week Perspectives Workshops

Each summer, Fresno Pacific University offers AIMS one-week workshops on its campus in Fresno, California. AIMS Program Directors and highly qualified members of the AIMS National Leadership Network serve as instructors.

The AIMS Instructional Leadership Program

This is an AIMS staff-development program seeking to prepare facilitators for leadership roles in science/math education in their home districts or regions. Upon successful completion of the program, trained facilitators may become members of the AIMS Instructional Leadership Network, qualified to conduct AIMS workshops, teach AIMS in-service courses for college credit, and serve as AIMS consultants. Intensive training is provided in mathematics, science, process and thinking skills, workshop management, and other relevant topics.

College Credit and Grants

Those who participate in workshops may often qualify for college credit. If the workshop takes place on the campus of Fresno Pacific University, that institution may grant appropriate credit. If the workshop takes place off-campus, arrangements can sometimes be made for credit to be granted by another institution. In addition, the applicant's home school district is often willing to grant in-service or professional-development credit. Many educators who participate in AIMS workshops are recipients of various types of educational grants, either local or national. Nationally known foundations and funding agencies have long recognized the value of AIMS mathematics and science workshops to educators. The AIMS Education Foundation encourages educators interested in attending or hosting workshops to explore the possibilities suggested above. Although the Foundation strongly supports such interest, it reminds applicants that they have the primary responsibility for fulfilling *current* requirements.

For current information regarding the programs described above, please complete the following:

Information Request

Please send current information on the items checked:

___ *Basic Information Packet* on AIMS materials
___ *AIMS Instructional Leadership Program*
___ *AIMS One-Week Perspectives* workshops

___ *A Week with AIMS®* workshops
___ Hosting information for *A Day with AIMS®* workshops
___ Hosting information for *A Week with AIMS®* workshops

Name _____ Phone _____

Address _____

      Street   City   State   Zip

AIMS Program Publications

GRADES K-4 SERIES

Bats Incredible!
Brinca de Alegría Hacia la Primavera con las Matemáticas y Ciencias
Cáete de Gusto Hacia el Otoño con la Matemáticas y Ciencias
Cycles of Knowing and Growing
Fall Into Math and Science
Field Detectives
Glide Into Winter With Math and Science
Hardhatting in a Geo-World (Revised Edition, 1996)
Jaw Breakers and Heart Thumpers (Revised Edition, 1995)
Los Cincos Sentidos
Overhead and Underfoot (Revised Edition, 1994)
Patine al Invierno con Matemáticas y Ciencias
Popping With Power (Revised Edition, 1996)
Primariamente Física (Revised Edition, 1994)
Primarily Earth
Primariamente Plantas
Primarily Physics (Revised Edition, 1994)
Primarily Plants
Sense-able Science
Spring Into Math and Science
Under Construction

GRADES K-6 SERIES

Budding Botanist
Critters
El Botanista Principiante
Exploring Environments
Fabulous Fractions
Mostly Magnets
Ositos Nada Más
Primarily Bears
Principalmente Imanes
Water Precious Water

GRADES 5-9 SERIES

Actions with Fractions
Brick Layers
Brick Layers II
Conexiones Eléctricas
Down to Earth
Electrical Connections
Finding Your Bearings (Revised Edition, 1996)
Floaters and Sinkers (Revised Edition, 1995)
From Head to Toe
Fun With Foods
Gravity Rules!
Historical Connections in Mathematics, Volume I
Historical Connections in Mathematics, Volume II
Historical Connections in Mathematics, Volume III
Just for the Fun of It!
Machine Shop
Magnificent Microworld Adventures
Math + Science, A Solution
Off the Wall Science: A Poster Series Revisited
Our Wonderful World
Out of This World (Revised Edition, 1994)
Paper Square Geometry: The Mathematics of Origami
Pieces and Patterns, A Patchwork in Math and Science
Piezas y Diseños, un Mosaic de Matemáticas y Ciencias
Proportional Reasoning
Ray's Reflections
Soap Films and Bubbles
Spatial Visualization
The Sky's the Limit (Revised Edition, 1994)
The Amazing Circle, Volume 1
Through the Eyes of the Explorers:
   Minds-on Math & Mapping
What's Next, Volume 1
What's Next, Volume 2
What's Next, Volume 3

For further information write to:
AIMS Education Foundation • P.O. Box 8120 • Fresno, California 93747-8120
www.AIMSedu.org/ • Fax 559•255•6396

We invite you to subscribe to *AIMS*®!

Each issue of *AIMS*® contains a variety of material useful to educators at all grade levels. Feature articles of lasting value deal with topics such as mathematical or science concepts, curriculum, assessment, the teaching of process skills, and historical background. Several of the latest AIMS math/science investigations are always included, along with their reproducible activity sheets. As needs direct and space allows, various issues contain news of current developments, such as workshop schedules, activities of the AIMS Instructional Leadership Network, and announcements of upcoming publications.

AIMS® is published monthly, August through May. Subscriptions are on an annual basis only. A subscription entered at any time will begin with the next issue, but will also include the previous issues of that volume. Readers have preferred this arrangement because articles and activities within an annual volume are often interrelated.

Please note that an *AIMS*® subscription automatically includes duplication rights for one school site for all issues included in the subscription. Many schools build cost-effective library resources with their subscriptions.

YES! I am interested in subscribing to *AIMS*®.

Name _____   Home Phone _____

Address _____   City, State, Zip _____

Please send the following volumes (subject to availability):

_____	Volume VI	(1991-92)	$30.00	_____ Volume XI	(1996-97)	$30.00
_____	Volume VII	(1992-93)	$30.00	_____ Volume XII	(1997-98)	$30.00
_____	Volume VIII	(1993-94)	$30.00	_____ Volume XIII	(1998-99)	$30.00
_____	Volume IX	(1994-95)	$30.00	_____ Volume XIV	(1999-00)	$30.00
_____	Volume X	(1995-96)	$30.00	_____ Volume XV	(2000-01)	$30.00

_____ **Limited offer: Volumes XIV & XV (1999-2001) $55.00**
(Note: Prices may change without notice)

Check your method of payment:

☐ Check enclosed in the amount of $ _____

☐ Purchase order attached (Please include the P.O.#, the authorizing signature, and position of the authorizing person.)

☐ Credit Card   ☐ Visa   ☐ MasterCard   Amount $ _____

　　Card # _____   Expiration Date _____

　　Signature _____   Today's Date _____

Make checks payable to **AIMS Education Foundation.**
Mail to *AIMS*® Magazine, P.O. Box 8120, Fresno, CA 93747-8120.
Phone (559) 255-4094 or (888) 733-2467   FAX (559) 255-6396
AIMS Homepage: http://www.AIMSedu.org/

AIMS Duplication Rights Program

AIMS has received many requests from school districts for the purchase of unlimited duplication rights to AIMS materials. In response, the AIMS Education Foundation has formulated the program outlined below. There is a built-in flexibility which, we trust, will provide for those who use AIMS materials extensively to purchase such rights for either individual activities or entire books.

It is the goal of the AIMS Education Foundation to make its materials and programs available at reasonable cost. All income from the sale of publications and duplication rights is used to support AIMS programs; hence, strict adherence to regulations governing duplication is essential. Duplication of AIMS materials beyond limits set by copyright laws and those specified below is strictly forbidden.

Limited Duplication Rights

Any purchaser of an AIMS book may make up to *200 copies* of any activity in that book for use at *one school site*. Beyond that, rights must be purchased according to the appropriate category.

Unlimited Duplication Rights for Single Activities

An individual or school may purchase the right to make an unlimited number of copies of a single activity. The royalty is $5.00 per activity per school site.

Examples: 3 activities x 1 site x $5.00 = $15.00
9 activities x 3 sites x $5.00 = $135.00

Unlimited Duplication Rights for Entire Books

A school or district may purchase the right to make an unlimited number of copies of a single, *specified* book. The royalty is $20.00 per book per school site. This is in addition to the cost of the book.

Examples: 5 books x 1 site   x $20.00 = $100.00
12 books x 10 sites x $20.00 = $2400.00

Magazine/Newsletter Duplication Rights

Those who purchase *AIMS*® (magazine)/*Newsletter* are hereby granted permission to make up to 200 copies of any portion of it, provided these copies will be used for educational purposes.

Workshop Instructors' Duplication Rights

Workshop instructors may distribute to registered workshop participants a maximum of 100 copies of any article and/or 100 copies of no more than eight activities, provided these six conditions are met:

1. Since all AIMS activities are based upon the *AIMS Model of Mathematics* and the *AIMS Model of Learning*, leaders must include in their presentations an explanation of these two models.
2. Workshop instructors must relate the AIMS activities presented to these basic explanations of the AIMS philosophy of education.
3. The copyright notice must appear on all materials distributed.
4. Instructors must provide information enabling participants to order books and magazines from the Foundation.
5. Instructors must inform participants of their limited duplication rights as outlined below.
6. Only student pages may be duplicated.

Written permission must be obtained for duplication beyond the limits listed above. Additional royalty payments may be required.

Workshop Participants' Rights

Those enrolled in workshops in which AIMS student activity sheets are distributed may duplicate a maximum of 35 copies or enough to use the lessons one time with one class, whichever is less. Beyond that, rights must be purchased according to the appropriate category.

Application for Duplication Rights

The purchasing agency or individual must clearly specify the following:
1. Name, address, and telephone number
2. Titles of the books for Unlimited Duplication Rights contracts
3. Titles of activities for Unlimited Duplication Rights contracts
4. Names and addresses of school sites for which duplication rights are being purchased.

NOTE: Books to be duplicated must be purchased separately and are not included in the contract for Unlimited Duplication Rights.

The requested duplication rights are automatically authorized when proper payment is received, although a *Certificate of Duplication Rights* will be issued when the application is processed.

Address all correspondence to: **Contract Division**
AIMS Education Foundation
P.O. Box 8120
Fresno, CA 93747-8120

www.AIMSedu.org/
Fax 559•255•6396